Proceedings in Technology Transfer

Series Editors

The book series "Proceedings in Technology Transfer" provides a valuable platform for researchers, practitioners, and policymakers to share their experiences and insights related to technology transfer. Our aim is to facilitate knowledge exchange between academic researchers and industry professionals, accelerating the introduction of new technologies and driving economic growth.

Our series publishes books and conference proceedings that focus on original research findings, best practice recommendations, emerging trends, and technology evaluations. The series covers all topics which explore the transfer of technology in various industry sectors that require innovations and economic growth.

Join us in advancing the technology transfer and its impact on innovation. Publish your proceedings with us and offer insights and expertise to professionals in your field.

Kilian Gericke · Claudia Eckert · Vishal Singh ·
Srinivasan Venkataraman
Editors

Co-Evolution of Design Research and Design Practice

Implications of New Methodological Developments on Future Design Processes

Springer

Editors
Kilian Gericke
Universität Rostock
Rostock, Mecklenburg-Vorpommern, Germany

Claudia Eckert
School of Engineering and Innovation
The Open University
Milton Keynes, UK

Vishal Singh
Department of Design and Manufacturing
Indian Institute of Science
Bangalore, Karnataka, India

Srinivasan Venkataraman
Department of Design
Indian Institute of Technology Delhi
New Delhi, Delhi, India

ISSN 2948-2321 ISSN 2948-233X (electronic)
Proceedings in Technology Transfer
ISBN 978-3-032-16277-9 ISBN 978-3-032-16278-6 (eBook)
https://doi.org/10.1007/978-3-032-16278-6

This Springer imprint is published by the registered company Springer Nature Switzerland AG
The registered company address is: Gewerbestrasse 11, 6330 Cham, Switzerland

Contents

The Changing Practice of Engineering Design

Claudia Eckert[1](✉), Kilian Gericke[2], Vishal Singh[3], and Srinivasan Venkataraman[4]

[1] The Open University, Milton Keynes, UK
claudia.eckert@open.ac.uk
[2] University of Rostock, Rostock, Germany
kilian.gericke@uni-rostock.de
[3] IISC, Bangalore, India
[4] IIT Delhi, Delhi, India

The practice of engineering design has been going through rapid changes, especially in recent years. The changes are happening at an unprecedented speed, driven by multiple enabling factors, such as, widespread digitalization, resulting globalization and geo-political changes in the global supply chain, the rapid advancements in enabling technologies such as artificial intelligence and machine learning, wireless sensing and communication networks, and rapid and additive manufacturing. Thus, the engineering design practice is changing and co-evolving with the changes in society at large.

In a society that is increasingly technology and digitally mediated, the scope and impact of engineering design has expanded across the various dimensions of the cyber-physical space, such as the digital-physical, product-service, simple-complex, and standalone-ecosystem spectrums. Consequently, while simplification is sought in user interaction and user experience, product complexity is increasing by the day with the growing trend towards smart products, embedded systems, cyber-physical solutions, and product service systems.

While an increasing product complexity calls for greater integration across supply chains and across companies, a fragmenting and polarising world makes this coordination more difficult. Customers are no longer looking for isolated products that can be managed and maintained independently or combined with other products to achieve their goals. Instead, customers are looking for integrated solutions. On the one hand, this is leading to the development of product-service systems and a fair greater consideration and integration of services in product design, but it also requires a better coordination across companies and a standardisation of the interfaces across which they interact. At the same time, sustainability has become an increasingly important issue for all products, with legislators setting tough performance targets. There is an evident need and demand for sustainable performance of many products and regulatory nudge towards greater repairability and circularity. While all designers and engineers across the world must deal with sustainability challenges, the regulatory context across different countries become increasingly more varied, leaving global companies with the challenge of being competitive in different market contexts. This will require the development of new design theories, guidelines, methods, tools, interventions, etc. which when implemented will help quell the listed challenges and mitigate the pain points. The implementation may bring additional challenges, which can be mitigated in subsequent iterations. This transition from practice-to-theory-to-practice is important.

K. Gericke et al. (Eds.): DSC 2022/2023, PTT, pp. 1–5, 2026.
https://doi.org/10.1007/978-3-032-16278-6_1

The digital revolution has hit engineering designing in two ways. It has affected the design process by making a lot more information available from past designs, and given designers the ability to model and simulate in new ways. At the same time, the products need to be designed in a way that enables the collection of relevant data to monitor, maintain and adapt. This has led to a greater proliferation of different academic backgrounds coming into engineering design. This proliferation is happening at a time when engineering design has branched out across various sub-specializations, across different partners and locations in the supply chain, and more agile and flexible, to the point that many more engineers work from home. This spread and diversity has made the planning, management and coordination of design teams more challenging even though the computer support for this coordination has greatly improved. At the same time, new technologies, materials and manufacturing methods are being developed all the time and engineering designers need to respond to these developments.

These changes have hugely broadened the issues that need to be considered during the engineering design process and made the task of understanding, supporting and improving design processes much harder. The academic design research community tries to support industry in this transition and is responding with the development of new tools and methods. However, having traditionally been reactive to the industry needs, there has often been a time delay between problems occurring in practice and the response of the academia. The Design Process Special Interest Group (DP SIG) of The Design Society addresses this gap by facilitating a dialogue between industry and academia and addressing some of the fundamental questions that many companies and researchers need to answer to respond to the challenges outlined above.

The DP SIG aims to support industry in understanding, modelling and running design processes by bringing together a community of design researchers and interacting with industry by identifying research challenges in modelling and managing engineering processes and working together to resolve them. The SIG investigates strategies, approaches, methods, and tools for modelling, building up, optimising and managing design processes of complex engineering products.

The DP SIG of The Design Society is one of the oldest special interest groups. It started out as MMEP, modelling and managing engineering processes, reflecting the early 2000s' interests in planning and structuring product development processes. The SIG meets regularly at conferences and also runs individual workshops, like the Munich workshop, from which many of the chapters in this book have originated. As many Special Interest Groups in the Design Society it is largely focused on central and north-western Europe, but the members collaborate internationally and this book also includes contributions from a workshop at the International Conference on Research into Design - ICoRD'23, held at the Indian Institute of Science, Bengaluru, where Indian participants were also joined online by contributors from the earlier workshop.

The chapters of this book started as talks at two workshops that the SIG organised in 2022 and 2023. The first workshop took place at the Chair of Product Development and Lightweight Structures, Technical University of Munich in September, 2022, which brought together academics from different countries with industry experts. Many of the industry experts, have themselves worked on doctoral research in product development and focussed on the development of tools and methods in their theses. Most of the industrial authors of the chapters in this book have seen the challenges of engineering design research both as researchers in academia and as practitioners in industry. They have had to persuade their own colleagues to work on engineering design in a structured manner. Most of the academic participants work closely with industry either through cooperative projects or formal dialogues. The participants were given the following brief:

The Design Process Special Interest Group focuses on the theory and empirical studies of design processes. Structured processes are necessary to orchestrate the interplay of people and to assure that products are developed on time and on budget with appropriate approaches & procedures, methods, data, and tools.

We seek contributions that elaborate on the implications of current research on future design processes, as well as contributions of practitioners that help to understand current challenges that require further research. During the workshop, we aim to discuss the impact on design processes and interdependencies between evolving research results and practice.

Guiding questions for contributions and discussions can be:

- *How do new methods, technologies or other research results affect the design process?*
- *Do they lend themselves to a specific paradigm, i.e. stage-gate-based or agile project management?*
- *What challenges do companies face during the implementation of new approaches into their processes?*

Topics could include, but are not limited to: New methods for new technologies, New process paradigms, Process modelling, Coordination of activities in design processes.

The second workshop took place in the Department of Design and Manufacturing (erstwhile Centre for Product Design and Manufacturing) at the Indian Institute of Science, Bengaluru in India six months later. An open invitation was sent to researchers in India inviting them to submit abstracts on topics related to 'Co-evolution of design research and practice', following which the abstracts were screened for their suitability and the authors of the relevant abstracts were invited to present their work in a workshop held immediately after ICoRD'23. Following the workshop, these researchers were invited to submit book chapters. The Indian design community is currently going from strength to strength as India is becoming a more industrialized economy than before. Many western European companies now have a design and a R&D office in India, and Indian OEMs are increasingly engaging in the development of complex products. At the same time, the design research community in India has grown manifold in the recent years. For example, in the ICoRD series of conference the number of attendees has grown exponentially in the last five years, especially representing a wide variety of design research domains. Consequently, the discussions at the workshop in India complement the discussions – representing the European viewpoint – in Germany.

The first workshop largely focussed on the exchange between industry and academia as a large share of the participants were industry representatives while the second workshop was associated with an academic conference. This also accounts for the subject matter of the presentations. The chapters to emerge out of the discussions in the Munich workshop include Chapters 2 to 10.

The book starts with four chapters characterising the fundamental changes that product development practice is going through. Chapter 2 (ref) explores the dynamic co-evolution of tools, methods, and processes within product development ecosystems, emphasizing how systemic interdependencies and disruptive technologies like AI shape and transform design practices. Chapter 3 (ref) addressed the increasingly interdisciplinary nature of engineering design and the challenges that this poses for the use of tool and methods, when so many participants come from different traditions with a diverse understanding of the product and development process. Chapter 4 (ref) discusses how digitisation is affecting engineering design processes in terms of the maturity of the digital models required at particular points in the process, the flexibility to restructure and replan processes and the ability to communicate electronically. Chapter 5 (ref) continues the themes of changing processes, by comparing product development and project management in agile and conventional practices and investigating how the two paradigms can be combined.

The introduction of agile process methods, which originated in software design, into the design of physical products is an important theme in industry, in particular in the German speaking countries. Two papers report on case studies of agile development. Chapter 6 (ref) compare the practices in two companies designing classical engineering products and analyse the importance of leadership and training for agile processes. Chapter 7 (ref) discusses the application of agile methods, in particular scrum to ship building. The paper proposes an agile workflow for this highly regulated sector. Chapter 8 (ref) presents the findings of a series of studies and investigates how a strategy originating from software can be applied to physical products. It points out, that the high level of interconnectivity of physical products requires careful thinking through agile work processes and thoughtful management.

Chapter 9 (ref) address another shift from the design of products to the design of product service systems, and find that this transition is also affected by the general challenges of embedded new approaches into industrial practice, and highlight the importance of embracing data centred approaches in integrating services and products. Chapter 10 (ref) also highlights the importance of good digital records and company resources, but in the context of model-based system engineering in battery development. The paper shows how a company has embraced MBSE and AI methods to retrieve relevant information from company databases, to help engineers handle to complexity and information volume involved in designing different versions of car batteries.

The papers from the Indian workshop move away from the challenges of designing and developing complex products in industry to address more theoretical aspects of engineering design. Chapter 11 (ref) looks at the Co-evolution of problem and solution space in the context of generative design arising from the need to represent the design problem space in a suitable way. The chapter also discusses the types of knowledge–domain knowledge, computational knowledge, and representational knowledge, that are required in generative design process, and how the relevance of these co-evolves during the process. Chapter 12 (ref) compares human-centric vs design-centric sources of pain points, needs and requirements and argue that while a design-centric analysis provides the broader picture, a human centred view is required to identify nuances. Chapter 13 (ref) also picks up on the co-evolution of problem and solution spaces through different thinking paradigms: design thinking, systems thinking and computational thinking, and argues that these different thinking styles are intertwined and need to be applied in conjunction. The paper also identifies the methodological implications of connecting these different thinking styles.

The last two chapters address the challenges and opportunities these changes bring for engineering design education. Chapter 14 (ref) carries on the theme of co-evolution. This time of design research and design practice, which must be reflected in design education. It also argues for the need to see design education in a more cross disciplinary way and the societal challenges like sustainability. Sustainability education is the focus of Chapter 15 (ref), which identified a broad range of educational practices with regards to sustainable product development across 118 German higher educational institutions.

Co-evolution of Tools and Methods in Product Development Ecosystems

Kilian Gericke[1(✉)] and Claudia Eckert[2]

[1] University of Rostock, Rostock, Germany
kilian.gericke@uni-rostock.de
[2] The Open University, Milton Keynes, UK

Abstract. Product development can be understood as a dynamic socio-technical ecosystem comprising interdependent elements such as products, people, processes, methods, tools, ICT environments and their contexts. This ecosystem perspective emphasizes the evolutionary and adaptive nature of product development environments, where change is not isolated but propagated through complex networks of actors, processes, tools and infrastructures. Influences from regulatory transformations, sustainability demands, and advances in digital technologies continuously reshape these ecosystems, requiring an integrated, systemic understanding to navigate and manage their evolution effectively.

Within this framework, methods and tools are not static instruments but co-evolving entities embedded in a larger system of development practices. Their successful implementation depends on coherence with existing processes and compatibility with the knowledge and competencies of stakeholders. New tools often emerge from technological innovation—particularly in artificial intelligence and data analytics—and in turn reshape the methods that support development activities. When these changes are not aligned or are poorly integrated, they can generate disruptions analogous to ecological imbalances, leading to inefficiencies or resistance to adoption.

The increasing complexity of products, including cyber-physical and product-service systems, demands interdisciplinary collaboration and context-specific integration of practices. Disciplines such as mechanical engineering, computer science, and service design bring distinct approaches and expectations, making harmonization of processes and terminology essential. Organizational challenges, such as varying digital competencies and generational shifts in work habits, further contribute to the dynamic character of development ecosystems.

Artificial intelligence represents a potentially disruptive force within these ecosystems. While current applications primarily automate repetitive or data-intensive tasks, future systems are expected to support decision-making, contribute creatively, and operate as learning entities within development teams. Such transformation requires rethinking roles, workflows, and even the tacit knowledge that underpins engineering expertise. The interplay between human and machine intelligence is poised to redefine not only how products are developed but also how innovation processes are conceived and managed.

Effective adaptation to these changes depends on maintaining the internal coherence of the product development ecosystem. Tools and methods must evolve in concert, and their interdependencies must be understood and respected. Changes

K. Gericke et al. (Eds.): DSC 2022/2023, PTT, pp. 6–19, 2026.
https://doi.org/10.1007/978-3-032-16278-6_2

in one element often necessitate adjustments in others; otherwise, benefits may not materialise, or unintended consequences may arise. Ecosystem resilience depends on recognizing these interconnections and managing change systematically rather than through isolated interventions.

Product development is undergoing a paradigm shift that calls for continuous reassessment of practices, tools, and frameworks. Understanding the co-evolution of tools and methods within evolving ecosystems is critical to supporting innovation, improving resilience, and sustaining competitiveness in a rapidly changing environment. A robust response to the ongoing transformation requires interdisciplinary research capable of addressing the socio-technical, organizational, and technological dimensions of change.

1 Introduction

Product development is a socio-technical endeavour, as technological innovations are deeply embedded in social and geopolitical contexts. As social and political challenges become more pronounced, it becomes clear that product development cannot be approached solely from a technical perspective. Rather, an integrated approach is required that takes the complex interplay between technological solutions and the social environment in which they are used into account. In this chapter, product development is presented as an interconnected, context-dependent ecosystem. This concept is used to discuss evolutionary change and the impact of disruptive change, as well as the effects of change propagating through the ecosystem. Thinking of product development holistically as an ecosystem helps organisations to develop strategies to become more resilient to rapid changes that could affect different aspects of the product development ecosystem.

2 The Concept of an Ecosystem

The concept of an ecosystem is used in many different fields originating from biology. Seminal work on ecosystems was carried in the 1950s. Odum (1956) thought of the net ecosystem metabolism (NET), as the balance between the production and consumption of organic matter in a natural system; and thereby provided a means of comparing ecosystems over time and with each other. Ecosystems of any kind exhibit evolution, variability, adaptability and selection. They are affected by path dependency, i.e. previous choices or events affecting those that come afterwards. Path dependent systems display lock-in, self-reinforcing, increasing returns and positive feedback (see Levin et al. (2012) for a discussion of path dependency and climate change decisions).

The concept has gained in prominence in the management of innovation and technology community since the 1990s, which sees ecosystems as a form of multi-actor networks, centring around the business ecosystem, but involving also users, investors and policy makers. It has looked at ecosystems from four different perspectives (see Tsujimoto et al. 2018):*industrial ecology*, focusing on the optimisation, sustainability and symbiosis of the industrial ecosystems; *business ecosystems*, which look at organisational boundaries and the identification of business niches; *product platform management*, which looks at the stability and evolvability of the products; and *multi-actor networks*, which emphasise the embeddedness, resilience and evolvability of the systems of actors. They define the objective of an ecosystem as "*To provide a product/service system, an historically self-organized or managerially designed multilayer social network consists of actors that have different attributes, decision principles, and beliefs.*" (Tsujimoto et al. 2018). Within such a system, it is possible to identity causal relations and therefore levers for improvement. Each actor has their own attributes, experiences and believes arising from own past experiences and present goals.

Going back to the work of Odum (1956), Nielson (2007) addresses the relevance of the concept of an ecosystem for an economists view on sustainable development by comparing natural and cultural systems. He focusses on 10 key characteristics:

1. Complexity – components and interconnectedness;
2. Evolution – adaptation and selection;
3. Compartmentalisation – quantity and quality;
4. Flows and processes – quantity and quality;
5. Feedbacks and controls;
6. Cycling – cycles, cycling, and recycling;
7. Network properties – synergism and utility;
8. Organisation and hierarchy;
9. Diversity;
10. Open systems and dissipation.

Nielson cautions against the use of ecosystems as a loose metaphor and also highlights the utility of the analysis. These concepts are also relevant to the lifecycle of method ecosystems (Gericke, Eckert, Campean, P John Clarkson, *et al.* 2020a, b).

Ecosystems have distinct lifecycles. According to Moore (1993, 1996) an ecosystem progresses through four phases - birth, expansion, leadership, and self-renewal (or death). The innovation ecosystem research is interested in the genesis of ecosystems, i.e. how an ecosystem grows around one or many innovations and how these innovations are sustained in these ecosystems. We can think of methods as a distinct type of innovation, which similarly need an ecosystem to flourish. An innovation ecosystem can be thought of as the evolving set of actors, activities, and artefacts, and the institutions and relations, including complementary and substitute relations, that are important for the innovative performance of an actor or a population of actors (Granstrand and Holgersson 2020). According to Li-Ying et al. (2022) this definition highlights several important aspects of innovation ecosystems: they emerge within the context of an organisation and shape this organisation, they are shaped by the actors in the ecosystem, they only flourish in conjunction with entrepreneurship so that value can be generated. To understand how ecosystems come into being Dedehayir et al. (2018) conducted a systematic literature review and have identified four distinct classes of roles, which become prominent at different times: "leadership roles ('ecosystem leader' and 'dominator'), direct value creation roles ('supplier', 'assembler', 'complementor', and 'user'), value creation support roles ('expert' and 'champion'), and entrepreneurial ecosystem roles ('entrepreneur', 'sponsor', and 'regulator') - and defined in terms of the specific activities they carry out during ecosystem birth". Both research on innovation and organisation ecosystems (e.g. Talmar et al. (2020)) and the natural ecosystems agree that an ecosystem needs to be actively managed (e.g. Meraj et al. (2022)). One way to support and manage entrepreneurial ecosystems is by providing coherent digital infrastructure (Endres, Huesig and Pesch 2022).

3 Product Development Ecosystems

The product, the people involved, the processes, methods and ICT environment embedded in an organisation can be described as a product development ecosystem (PD ecosystem) (see Fig. 1). This view on the product development ecosystem focuses on the company developing the product. Other system boundaries would be drawn for different types of projects. The specific elements of an ecosystem that interact vary along the lifecycle of the product from its inception to its end of life. For example, experts with different backgrounds including manufacturing, operation or maintenance as well as different user representatives will interact. The design of products often involves multiple companies, which requires an interaction of different ecosystems.

PD ecosystems are context-specific (Hales and Gooch 2004; Gericke, Meißner and Paetzold 2013), characterised by an individual combination of the individual elements:

- the specific *products* or product families (including their underlying technologies and business model);
- the *people* who are involved in product development as project participants with their specific knowledge and specific skills and their specific roles/responsibilities and other project stakeholders along the product lifecycle;
- the *process* or the specifically necessary activities and the approach for their management;
- the *methods* used for designing and validating the product and management of processes;
- the *ICT environment* that supports the process as a whole or individual activities.

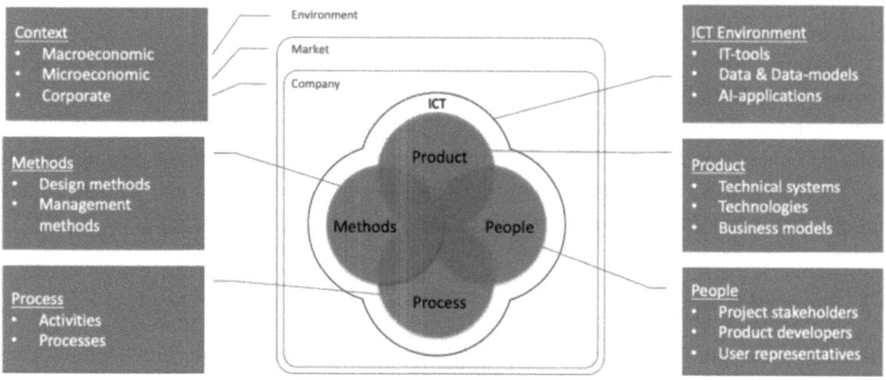

Fig. 1. Product Development Ecosystem(s)

An ecosystem is not static, but a dynamic system in which elements are added, removed or adapted when there is a need (Gericke et al. 2020a, b). Product development ecosystems must be resilient – and many are. However, like natural ecosystems, they can be sensitive to change.

4 Changes Affecting Product Development Ecosystems

The practice of product development is undergoing fundamental changes (Eckert and Isaksson no date; Eckert et al. 2019). New technologies and paradigms determine the development of future products (Rößler and Gericke 2022). New and changing development goals reflect the increased importance of sustainability, resilience and digitalisation (Hallstedt, Isaksson and Öhrwall Rönnbäck 2020).

The following section will illustrate a variety of changes that will affect product development ecosystems. The changes described are only examples and do not provide an exhaustive overview. These changes originate from the context, from within the PD ecosystem or from enabling technologies. Overall, we see several conflicting trends that will probably affect product development ecosystems. These exemplary trends affect sectors differently, thus different product development ecosystems are facing different challenges and need different solutions.

4.1 Changing Context

The context of product development is changing at both at macroeconomic and microeconomic scale. New regulatory and legal frameworks affect companies, products and related business models. Geopolitical changes affect companies, their value creation chains and access to markets. Effects of climate change become apparent leading to changed societal awareness resulting in an increased awareness for the benefits of a circular economy (Asgari and Asgari 2021; Kevin van Langen et al. 2021; Gonella et al. 2024). New technologies allow solutions that affect our society (Paul Tiwari 2022). The servitisation of solutions and changing societal value systems are leading to a shift in the dominance of sectors (Capello and Lenzi 2021).

Shifting geopolitics leads to deglobalisation in certain sectors, as countries decide to onshore or friendshore particular production capacities and raise import barriers for certain goods. This will significantly impact supply chains as suppliers from the same political sphere need to be identified. Trade, along with production and design, will be more influenced by international conflicts due to the use of sanctions, subsidies, and import taxes for political purposes (Wijaya and Jayasuriya, 2024; Gopinath et al., 2025).

4.2 Internal Changes

Products Change. To reach ever more demanding development goals, new technologies (sensor-integrated machine elements, artificial intelligence), energy sources (hydrogen, ammonia), materials (tailored materials), manufacturing processes and production systems (additive manufacturing, smart factories) are being used. In many cases, innovative products integrate not only physical artefacts but also electronic components, software or services. Product-service systems (PSS) and cyber-physical systems (CPS), some of which are operated in a coordinated network (Industry 4.0), are typical examples of modern, innovative solutions (Stark, Fresemann and Lindow 2019; Forschungsbeirat der Plattform Industrie 4.0/acatech - Deutsche Akademie der Technikwissenschaften 2022). The changing functionalities and the fulfilment of new ambitious development goals are usually accompanied by an enormous increase in product complexity and the integration of digital solution elements, as well as an increase in the amount and variety of data (Tomiyama et al. 2019). The changing nature of products also leads to a greater need for interdisciplinary work, which is discussed in more detail in chapter 3 in this book (Eckert, Muschik and Thau 2025).

People Change. Societal changes such as demographic change, urbanisation and digitalisation influence product development in various ways. The expectations of customers and users are constantly changing and are becoming increasingly individualised for some products, while other products interact with many users as well as other products and need simple to use and standardised solutions. The digital competences of young people develop rapidly, increasing the differences in skills and attitudes between different age cohorts (Kiser and Washington 2015; Bachmann et al. 2021). However, people influence product development not only as customers and users, but also directly as participants in the product development process as managers, developers, designers, production planners, etc., so that demographic developments and cohort-specific skills and different knowledge change the way people work together in product development.

Processes Change. The development of multi-technology products such as PSS and CPS requires specialised knowledge and skills from various disciplines. Products that used to be developed primarily by mechanical engineers are now developed by a team of experts from mechanical engineering, computer science, electrical engineering, service development and other disciplines. Each of these disciplines has developed its own approaches to developing new solutions and its own approaches to managing projects. A classic example is the introduction of agile approaches from software into mechanical engineering, where conventional approaches such as Stage Gate and agile approaches such as SCRUM are used together in interdisciplinary teams (Atzberger and Paetzold 2019). The current paradigm shift brings with it many challenges that will require context-specific answers and solutions. On the one hand, the challenges are characterised by the physicality of mechanical artefacts and the permanence of development decisions on development and manufacturing processes. This means that, unlike with software, changes to the artefact cannot simply be implemented afterwards. On the other hand, the challenges are also characterised by discipline-specific development cultures. This means that communities of practice often have established ways of working and are resistant to change, even if alignment with another community of practice (Wenger, 1998) is necessary. This can be observed, for example, in the cooperation between hardware and software development (Atzberger and Paetzold 2019; Peterson and Summers 2021).

Methods and Tools Change. Product developers use a variety of methods and ICT tools for the range of activities that need to be carried out during a product development project (Gericke, Kramer and Roschuni 2016). These pursue the goal of improving the effectiveness and efficiency of developers or enable activities that would otherwise not be possible and thus make an important contribution to the success of the project (Gericke, Eckert and Stacey 2017). Modern development methods and development tools are usually software-based. Current developments in artificial intelligence are already having an impact on available development tools. At the same time, there has been an enormous increase in research activities on the topic of AI-supported development methods. The progress made so far in the development of AI-based methods and tools is still rather incremental in nature and focuses on improved support for individual data-intensive and often repetitive development activities (Jonuschies et al. 2025). However, the forecast for further development progress of AI-based applications suggests that a stronger, fundamental change in development tools and development methods can be expected in the future (Verganti, Vendraminelli and Iansiti 2020).

These changes do not happen in isolation but influence each other. For example, as the functionality and technology of a product changes, people with different skills need to come into the process, who in turn bring their own understanding. This means that the people and methods need to be integrated into an ecosystem in a controlled way, otherwise elements start competing, for example over the meaning of terminology, which can lead to mistakes and delays.

4.3 Changes to the Enabling Technology

Developments in product development take place in the context of wider societal and technical trends (Eckert and Isaksson, no date).

Data. Product development is evolving as data collection capabilities continue to expand. With the ability to gather significantly more data, it is becoming crucial to integrate data design into the product development process. The challenge is to ensure that we collect the right data without exceeding what is necessary, balancing efficiency and purpose. As a result, data architecture design is set to become a major focus within the community, shaping how products are conceptualized and built. However, this shift also brings challenges, including concerns about privacy and the substantial energy required to store and manage large volumes of data. The same item of data might be of interest to multiple actors across the supply chains and the governance and access of data across multiple and potentially competing actors will require active management. Addressing these issues will be essential to making data-driven product development both sustainable and responsible.

The handling of large amounts of data is profoundly affected by the continuous improvement of artificial intelligence (AI). Future AI systems will not only analyse data, but also provide real-time, contextual recommendations based on comprehensive data models. This empowers product developers to identify trends at an early stage and secure a competitive advantage by responding flexibly to changing market conditions. However, these potentials rely on the availability and accessibility of appropriately managed data, which in practice is a major challenge.

Simulation and Analysis. Simulation and data analysis in engineering design are evolving rapidly due to advancements in computational power, data integration, and AI. Simulation techniques are now integrated into the early stages of the design process, allowing engineers to explore a greater number of alternatives and identify potential issues at an earlier stage. Real-time and interactive simulations, powered by GPU acceleration and cloud computing, provide instant feedback, thereby making the design process more dynamic. Furthermore, multiphysics simulations enable comprehensive analyses of complex systems, while cloud-based platforms facilitate global collaboration.

AI and machine learning enhance these simulations by predicting outcomes and optimizing designs based on data from previous projects. The emergence of digital twins allows for the creation of virtual replicas of physical systems that can be updated in real-time, thus supporting performance monitoring and maintenance throughout a product's lifecycle. To make sense of the results of simulation and analysis the engineers need to maintain an awareness of the quality of the data that is used in the calculations.

Automated optimization and uncertainty analysis have the potential to ensure robust and efficient designs provided the engineers understand the use situation and data quality, while sustainability assessments promote environmentally conscious engineering practices.

Artificial Intelligence in Product Development. Continuous change is a characteristic of the product development environment (Gericke and Eckert 2015). In contrast to many changes in the past, the use of artificial intelligence in product development has the potential to trigger disruptive change (Stark, Brandenburg and Lindow 2021). Previous evolutionary strategies for transforming PD ecosystems will not be sufficient to ensure successful transformation and prepare for current and future challenges.

The integration of AI into products and product development processes promises an even more far-reaching transformation in the future than is currently apparent. At present, AI is impressive in its ability to automate routine tasks, but in the future, it might also take over tasks that we now see as demanding and creative processes. This might make it possible to further reduce development times and to create products of higher precision and quality in less time. However, in the past the potential for reduced product development times has been taken up by increased product complexity.

A fundamental change will be in the role of AI itself: from a supporting tool to a fully-fledged team member. Future AI systems will not only perform tasks, but will also actively participate in decision-making, contribute new ideas and act as continuously learning partners in interdisciplinary teams (Verganti, Vendraminelli and Iansiti 2020). This will fundamentally reshape the way humans and machines work together and promote new forms of innovation culture.

The future development of AI will bring about a profound change in product development through its ability to take on more demanding tasks, generate innovative solutions and act as a creative player.

As a result, we are observing a transformation.

Product Development Ecosystems Change. The changes in context, products, people, processes, methods, tools and enabling technologies outlined above are subject to strong interactions (Lunnemann et al. 2017; Bender and Gericke 2021; Paetzold 2022). The resulting effects are mutually reinforcing in some cases and require changes in the product development ecosystem. Unlike incremental changes, to which ecosystems can adapt in a self-regulating way, disruptive changes represent a fundamental challenge.

5 Implications for Design Research

The changes described above are not yet fully realised. Many of the described potentials will require to understand their implications on the PD ecosystem to avoid negative effects. For example, historically, anything that could shorten development time led to rebound effects, such as more complex products that required a similar development time. Avoiding that is a fundamental challenge in its own right. Other challenges might stem from the shifting expertise of designers. They might lose skills that used to be required for routine tasks that will be taken over by more powerful tools; thereby the designers might use the tacit skills required to recognise correct or interesting solutions. It is essential to understand these effects and to assess whether they are relevant for the individual product development ecosystem before adopting new technologies, instead of facing unwanted consequences afterwards.

5.1 Co-evolution with Enabling Technology

Product development ecosystems and their contexts are evolving alongside with enabling technologies. This ongoing change in design practices creates a constant need for further investigation and improvement of methods, tools and processes. As new technologies emerge, they influence the way products are conceived, developed, and brought to market. This necessitates continuous adaptation and refinement of the existing frameworks and systems to keep pace with technological advancements.

For example, AI will impact the way product developers work individually and in teams, at the level of individual tasks and at the level of the overall process. It will impact all aspects of development practice. At the same time, this means that the scientific foundations of current product development methods, product development processes and ICT tools must be further developed in order to provide a reliable and effective basis for the development of suitable PD ecosystems.

5.2 Co-evolution of Methods

Design methods are elements of the product development ecosystem. While they are dependent on the other elements, they also depend on other design methods. Changes to the enabling technologies might also lead to new or changed design methods.

Changes to methods or the introduction of new methods go beyond replacing an old method or providing a support for activities that were executed without systematic support before. Such changes might affect other methods in a method ecosystem (Gericke, Eckert, Campean, P. John Clarkson, *et al.* 2020a, b) which is a smaller ecosystem within the product development ecosystem that needs to be understood and nurtured. Methods require certain inputs (information, resources, etc.) and will help to create certain outputs. If methods change, the interdependencies between methods might require changes to dependent methods or the expected benefits of using new methods will not unfold. Maintaining internal coherence is critical for a successful evolution of product development ecosystems. Thinking of methods as part of an ecosystem also explains why the introduction of plausible methods can be ineffective or counterproductive, as they might not be able to interact with other methods or provide competing results.

5.3 Co-evolution of Methods and Tools

Since many modern design methods and tools are usually software-based, they need to evolve together. Often, new developments in generic tools and enabling technologies trigger the evolution of design methods or allow the development of entirely new methods. With the advent of AI technologies, it can be assumed that a whole new generation of design methods will be developed along with tools that facilitate the use of these new methods.

5.4 Co-evolution of Tools and Processes

Product development processes are complex, and their structure is affected by the nature of the product and the skills of the designers, but also other elements in the product development ecosystem. The interrelationship between processes and tools is important. Design processes often reflect the capabilities and functionalities of the tools that are used for product development and management. Many tools are used in multiple contexts and product development ecosystems and thus not tailored to the particular needs of a specific ecosystem. This dependency means that the evolution of processes is often constrained by tools.

5.5 Co-evolution in an Ecosystem

The evolution of product development ecosystems is the result of the co-evolution of its constituent elements as well as external pressures that force the ecosystem to adapt sometimes faster than it can. The analogy of an ecosystem is useful for understanding the interconnected dependencies and the plurality of suitable alternative instances that will work in similar contexts.

The complexity of the interdependencies between the different elements of PD ecosystems requires a systemic understanding of the organisation and hierarchy of its elements and their relationships. It is important to understand that the network like properties of PD ecosystems will unfold dynamically, thus the interaction between stakeholders with another and with artefacts created during the PD process will cause feedback cycles that need to be understood and managed.

While it is important to have a systemic understanding of PD ecosystems on an abstract level it is equally important to understand the diversity of real PD ecosystems and the resulting challenges when aiming to adapt existing eco-systems. It is the complexity of context-specific dynamic and evolving interdependencies which render quick fixes often inappropriate. Ecosystems will only flourish if all of its elements interact successfully without dominance of one particular element. The ecosystems of different products and organisation share many abstract characteristics and problems, making it possible to transfer insights from one to the other. They also often intersect. However, there are differences between the different ecosystems that lead to the success of some and the ultimate demise of others.

6 Research Needs

Companies that are developing products are facing a variety of changes. If PD ecosystems are not adapted quickly enough, this will result in a competitive disadvantage for industry. Design research must now act in a concerted and comprehensive manner to facilitate companies in the transformation of PD ecosystems.

The transformation of existing ecosystems represents an enormous challenge for all stakeholders. Continuous improvement and adaptation of product development ecosystems requires monitoring and forecasting changes in each individual area. Understanding changes requires an understanding of how they will affect each other and how these changes will affect the efficacy and performance of the PD ecosystem as a whole. This systemic understanding is critical for the development of new or improved solutions aimed at supporting practitioners and for the selection and implementation of suitable solutions. A lacking understanding by researchers might result in solutions that do not fit into real PD contexts while a lacking understanding by practitioners might result in change efforts within PD ecosystems that do not create the expected benefit or even disturb an ecosystem.

The necessary research includes the investigation of the need for method adaptation and method development against the background of the previously outlined changes in PD ecosystems. This requires a profound understanding of the necessary knowledge and skills in product development, which is currently not available in sufficient quality and timeliness. A thorough understanding of contemporary engineering principles and techniques is essential for the planning, development, and evaluation of future methods, tools and processes in product development.

Besides effects of steady changes within each area of the PD ecosystem and their interdependencies, it is necessary to research effects of disruptive changes and how to manage the adaptation of PD ecosystems in such situations. This can be illustrated with the rise of AI. The increased use of AI in product development will permanently change the methods and processes, as well as computer-based tools and data models used in engineering. The resulting changes are likely to be disruptive as not only individual methods or tools will be changed, but also the enabling technologies affecting methods, tools, processes and the people involved in product development. The potential of AI for product development can only be fully exploited if the organisational and socio-technical framework conditions are taken into account (Stowasser and Neuburger 2022). It is essential to anticipate how the new division of labour between humans and AI can be optimally designed (André and Bauer 2021). This division of labour requires not only technological adjustments, but also a redesign of work organisation and work processes. It is therefore necessary to identify the interactions between technological issues and socio-technical requirements, and to offer solutions for how companies can use AI to create value.

Changes affecting elements of the PD ecosystem can lead to evolutionary transformation or disruptive transformation of the ecosystem. In order to cope with the resulting challenges and to provide valuable support to industry, academia must take an interdisciplinary research approach involving experts e.g. from computer science, data science, social science, management science and philosophy in technology. Design researchers can play a central role in shaping such interdisciplinary research efforts and in integrating results into actionable support for design practitioners.

In further work, it is well worth to explore the analogies of product development ecosystems to biological ecosystems in more details to explain the mechanism of introducing new methods or tools similar to introducing invasive species.

References

André, E., Bauer, W.: Kompetenzentwicklung für KI - Veränderungen, Bedarfe und Handlungsoptionen. München (2021). https://doi.org/10.48669/pls_2021-2

Asgari, A., Asgari, R.: How circular economy transforms business models in a transition towards circular ecosystem: the barriers and incentives. Sustain. Product. Consumpt. **28**, 566–579 (2021). https://doi.org/10.1016/j.spc.2021.06.020

Atzberger, A., Paetzold, K.: Current challenges of agile hardware development: what are still the pain points nowadays?' Proc. Des. Soc.: Int. Conf. Eng. Des. **1**(1), 2209–2218 (2019). https://doi.org/10.1017/dsi.2019.227

Bachmann, R., Hertweck, F., Kamb, R., Lehner, J., Niederstadt, M., Rulff, C.: Digitale Kompetenzen in Deutschland-eine Bestandsaufnahme. Essen (2021). www.rwi-essen.de

Bender, B., Gericke, K. (eds.): Pahl/Beitz Konstruktionslehre, vol. 9. Springer Berlin Heidelberg (2021). https://doi.org/10.1007/978-3-662-57303-7

Capello, R., Lenzi, C.: Industry 4.0 and servitisation: regional patterns of 4.0 technological transformations in Europe. Technol. Forecast. Soc. Change **173**, 121164 (2021). https://doi.org/10.1016/j.techfore.2021.121164

Dedehayir, O., Mäkinen, S.J., Roland Ortt, J.: 'Roles during innovation ecosystem genesis: a literature review. Technol. Forecast. Soc. Change **136**, 18–29 (2018). https://doi.org/10.1016/j.techfore.2016.11.028

Eckert, C., Isaksson, O.: Co-evolution of design research and industrial development: empirical insights from European manufacturing companies. J. Eng. Des. [Preprint] (no date)

Eckert, C., Isaksson, O., Hallstedt, S., Malmqvist, J., Öhrwall Rönnbäck, A., Panarotto, M.: Industry Trends to 2040. Proc. Des. Soc.: Int. Conf. Eng. Des. **1**(1), 2121–2128 (2019). https://doi.org/10.1017/dsi.2019.218

Eckert, C., Muschik, S., Thau, S.: Engineering methods in multidisciplinary processes. In: Gericke, K., Eckert, C., Singh, V., Venkataraman, S. (eds.) Co-Evolution of Design Research and Design Practice. Springer (2025)

Endres, H., Huesig, S., Pesch, R.: Digital innovation management for entrepreneurial ecosystems: services and functionalities as drivers of innovation management software adoption. Rev. Manager. Sci. **16**(1), 135–156 (2022). https://doi.org/10.1007/s11846-021-00441-4

Forschungsbeirat der Plattform Industrie 4.0/acatech - Deutsche Akademie der Technikwissenschaften. Themenfelder Industrie 4.0 - Forschungs- und Entwicklungsbedarfe zur erfolgreichen Umsetzung von Industrie 4.0. München (2022). https://doi.org/10.48669/fb40_2022-04

Gericke, K., et al.: Supporting designers: moving from method menagerie to method ecosystem. Des. Sci. **6**, e21 (2020a). https://doi.org/10.1017/dsj.2020.21

Gericke, K., et al.: Supporting designers: moving from method menagerie to method ecosystem. Des. Sci. **6** (2020b). https://doi.org/10.1017/dsj.2020.21

Gericke, K., Eckert, C., Stacey, M.: What do we need to say about a design method?. In: Proceedings of the International Conference on Engineering Design, ICED. The Design Society, Glasgow (2017)

Gericke, K., Kramer, J., Roschuni, C.: an exploratory study of the discovery and selection of design methods in practice. J. Mech. Des. **138**(10), 101109 (2016). https://doi.org/10.1115/1.4034088

Gericke, K., Meißner, M., Paetzold, K.: Understanding the context of product development. In: Lindemann, U., Venkataramen, S., Kim, Y., Lee, S., Cantamessa, M., Yannou, B. (eds.) Proceedings of the International Conference on Engineering Design, ICED. The Design Society, Glasgow (2013)

Gonella, J. dos S.L., Godinho Filho, M., Ganga, G.M.D., Latan, H., Chiappetta Jabbour, C.J.: A behavioral perspective on circular economy awareness: the moderating role of social influence and psychological barriers. J. Clean. Product. **441**, 141062 (2024). https://doi.org/10.1016/j.jclepro.2024.141062

Gopinath, G., Gourinchas, P.-O., Presbitero, A.F., Topalova, P.: 'Changing global linkages: a new Cold War? J. Int. Econ. **153**, 104042 (2025). https://doi.org/10.1016/j.jinteco.2024.104042

Granstrand, O., Holgersson, M.: Innovation ecosystems: a conceptual review and a new definition. Technovation, 90–91, 102098 (2020). https://doi.org/10.1016/j.technovation.2019.102098

Hales, C., Gooch, S.: Managing engineering design, 2nd edn. Springer-Verlag, London (2004)

Hallstedt, S., Isaksson, O., Öhrwall Rönnbäck, A.: The need for new product development capabilities from digitalization, sustainability, and servitization trends. Sustainability **12**(23), 10222 (2020). https://doi.org/10.3390/su122310222

Jonuschies, I., Siewert, L., Michaelsen, E., Gericke, K.: Mapping AI applications in design: Support and activities across the product development process. In: Proceedings of the Design Society. Cambridge University Press (2025)

Kevin van Langen, S., Vassillo, C., Ghisellini, P., Restaino, D., Passaro, R., Ulgiati, S.: Promoting circular economy transition: a study about perceptions and awareness by different stakeholders groups. J. Clean. Product. **316**, 128166 (2021). https://doi.org/10.1016/j.jclepro.2021.128166

Kiser, A.I.T., Washington, R.: the information gap amongst the generations and the implications for organizations. Int. J. Dig. Literacy Dig. Compet. **6**(2), 36–63 (2015). https://doi.org/10.4018/IJDLDC.2015040103

Levin, K., Cashore, B., Bernstein, S., Auld, G.: Overcoming the tragedy of super wicked problems: constraining our future selves to ameliorate global climate change. Policy Sci. **45**(2), 123–152 (2012). https://doi.org/10.1007/s11077-012-9151-0

Li-Ying, J., Sofka, W., Tuertscher, P.: Managing innovation ecosystems around Big Science Organizations. Technovation **116**, 102523 (2022). https://doi.org/10.1016/j.technovation.2022.102523

Lunnemann, P., Stark, R., Wang, W.M., Stark, R., Manteca, P.I.: 'Engineering activities — considering value creation from a holistic perspective. In: 2017 International Conference on Engineering, Technology and Innovation (ICE/ITMC), pp. 315–323. IEEE (2017). Available at: https://doi.org/10.1109/ICE.2017.8279904

Meraj, G., Singh, S.K., Kanga, S., Islam, Md.N.: Modeling on comparison of ecosystem services concepts, tools, methods and their ecological-economic implications: a review. Model. Earth Syst. Environ. **8**(1), 15–34 (2022). https://doi.org/10.1007/s40808-021-01131-6

Moore, J.F.: Predators and prey: a new ecology of competition. Harvard Bus. Rev. **71**(3), 75–86 (1993). http://www.ncbi.nlm.nih.gov/pubmed/10126156

Moore, J.F.: The death of competition : leadership and strategy in the age of business ecosystems. Harper Business, New York City (1996)

Nielsen, S.N.: What has modern ecosystem theory to offer to cleaner production, industrial ecology and society? The views of an ecologist. J. Clean. Product. **15**(17), 1639–1653 (2007). https://doi.org/10.1016/j.jclepro.2006.08.008
Odum, H.T.: Primary production in flowing waters1. Limnol. Oceanography **1**(2), 102–117 (1956). https://doi.org/10.4319/lo.1956.1.2.0102
OECD. Bericht zu Künstlicher Intelligenz in Deutschland. OECD (2024). https://doi.org/10.1787/8fd1bd9d-de
Paetzold, K.: Data and information flow design in product development. In: Design Methodology for Future Products, pp. 201–218. Springer International Publishing, Cham (2022). https://doi.org/10.1007/978-3-030-78368-6_11
Paul Tiwari, S.: The impact of new technologies on society: a blueprint for the future. RS Global Sp. z O.O (2022). https://doi.org/10.31435/rsglobal/053
Peterson, M., Summers, J.D.: Recommended methods supporting adoption of the agile philosophy for hardware development. In: Volume 6: 33rd International Conference on Design Theory and Methodology (DTM). American Society of Mechanical Engineers (2021). https://doi.org/10.1115/DETC2021-70621
Rößler, L., Gericke, K.: Analysing paradigms for managing product development: conventional, agile and hybrid approaches. In: Proceedings of the Design Society, pp. 263–272. Cambridge University Press (2022). https://doi.org/10.1017/pds.2022.28
Stark, R., Brandenburg, E., Lindow, K.: Characterization and application of assistance systems in digital engineering. CIRP Ann. **70**(1), 131–134 (2021). https://doi.org/10.1016/j.cirp.2021.04.061
Stark, R., Fresemann, C., Lindow, K.: Development and operation of Digital Twins for technical systems and services. CIRP Ann. **68**(1), 129–132 (2019). https://doi.org/10.1016/j.cirp.2019.04.024
Stowasser, S., Neuburger, R.:x Führung im Wandel: Herausforderungen und Chance durch KI. München (2019). https://doi.org/10.48669/pls_2022-4
Talmar, M., Walrave, B., Podoynitsyna, K.S., Holmström, J., Romme, A.G.L.: Mapping, analyzing and designing innovation ecosystems: the ecosystem pie model. Long Range Plan. **53**(4), 101850 (2020). https://doi.org/10.1016/j.lrp.2018.09.002
Tomiyama, T., Lutters, E., Stark, R., Abramovici, M.: Development capabilities for smart products. CIRP Ann. **68**(2), 727–750 (2019). https://doi.org/10.1016/j.cirp.2019.05.010
Tsujimoto, M., Kajikawa, Y., Tomita, J., Matsumoto, Y.: A review of the ecosystem concept—towards coherent ecosystem design. Technol. Forecast. Soc. Change **136**, 49–58 (2018). https://doi.org/10.1016/j.techfore.2017.06.032
Verganti, R., Vendraminelli, L., Iansiti, M.: Innovation and design in the age of artificial intelligence. J. Product Innov. Manage. **37**(3), 212–227 (2020). https://doi.org/10.1111/jpim.12523
Wenger, 1998. Wenger, E.: Communities of practice. Cambridge University Press (1998). https://doi.org/10.1017/CBO9780511803932
Wijaya, T., Jayasuriya, K.: Militarised neoliberalism and the reconstruction of the global political economy. New Political Econ. **29**(4), 546–559 (2024). https://doi.org/10.1080/13563467.2023.2298199

Engineering Methods in Multidisciplinary Processes

Claudia Eckert[1(✉)], Sabine Muschik[2], and Sebastian Thau[3]

[1] The Open University, Milton Keynes, UK
claudia.eckert@open.ac.uk
[2] Trumpf, Ditzingen, Germany
[3] precote – omniTECHNIK, Munich, Germany

Abstract. Engineering products increasingly involve the collaboration of engineers from many different disciplines from mechanical engineering to software engineering and systems engineering as well as non-engineering experts, like data scientists, natural scientists or mathematicians. Each discipline has its own terminology, tools and methods, as well as a different mindset and logic by which work is structured. At present the terminology, tools and methods used in an organization have often come from the original discipline of the product. This paper reports the practical challenges in managing engineering design processes where multiple disciplines are involved. It outlines the implications for the development of tools and methods to bridge the disciplinary boundaries. The use of methods is not unproblematic even within a single discipline, as many methods proposed by academia are too complex for industry and practitioners cherry pick the parts of methods they like. The understanding and use of methods is strongly influenced not only by the discipline, but the specialisation and its school of origine, as well as the company context, while some participants, like chemists, have no experience of engineering methods at all. Challenges in the uptake of methods of do not only come from the method itself, but also a divergent underlying understanding of the purpose of the product, the technical system and the process by which it is generated. The paper argues that to increase the uptake of methods in organisations an active effort is required to develop a shared understanding of the methods in general and the mind set behind a specific method as well as the details of the method.

1 Introduction

Engineering products increasingly involve the collaboration of engineers from many different disciplines from mechanical engineering to software engineering and systems engineering. While engineers are basically open to collaboration with other disciplines, engineers and engineering managers find this difficult in practice. Over a career many engineers become familiar with some of the disciplines they work with, but this takes time and stability that many modern engineering companies can't provide. Just as the best formular 1 drivers are those who have a deep technical understanding, an overlap in understanding between different fields is a critical success factor.

K. Gericke et al. (Eds.): DSC 2022/2023, PTT, pp. 20–35, 2026.
https://doi.org/10.1007/978-3-032-16278-6_3

Each discipline has its own terminology, tools and methods, as well as a different mindset and logic by which work is structured. Superimposing the methodologies of one discipline onto another can be challenging, and at worst undermine the professional integrity and self-understanding of a discipline. At present the terminology, tools and methods used in an organization have often come from the original discipline of the product. For example, if the product was once purely mechanical, the company was traditionally run by mechanical engineers who would have made the assumption that other disciplines fit into mechanical engineering ways of approaching the design. Collaboration becomes even more of an issue if it goes outside traditional engineering disciplines, for example to working with scientists or mathematicians. This paper reports the practical challenges in managing engineering design processes where multiple disciplines are involved. It outlines the implications for the development of tools and methods to bridge the disciplinary boundaries.

It is well recognized that communication and collaboration in engineering teams in difficult as individuals are influenced by their own expertise and experience. Larry Bucciarelli uses the term object worlds "to refer to the worlds of individual effort where an engineer, working for the most part alone, applies his or her expertise to particular tasks appropriate to his or her discipline. I claimed that different participants, with different competencies, skills, responsibilities and interests, inhabit different worlds. As such, while admittedly working on the same object of design, they see the object differently" (Bucciarelli 2002). These object worlds are both personal and deeply influenced by the engineers' disciplines as specific scientific and instrumental paradigms fix meaning. Bucciarelli (2002) argues that object worlds have their own languages, where people refer to the same element or behaviour of an object with different terms and conceptualization. He points out that these are "proper" languages, with clearly defined but different meanings, which can be translated. Unlike other forms of languages, people are often not aware that they are hearing a different language and therefore that translation is required. Bucciarelli sees this as one of the reasons why many design projects fail or at least do not deliver as hoped for. In the last 30 years, since the awareness of communication difficulties has been rising, steps have also been taken to ameliorate them and communication issues are seen as the manifestation of underlying problems (Maier et al. 2021). Many of these issues are aggravated when experts from different engineering disciplines or even from outside of engineering try to collaborate, as the examples presented in the paper illustrate. However, Bucciarelli (2002) also points out that "disciplinary perspective of the particular world and particular metaphors [can] enlighten and enliven the efforts of inhabitants".

This paper looks at the use of methods in industry in a multidisciplinary context; however the use of methods even within a single discipline is not unproblematic. Wallace (2011) points out that "methods tend to be too complex, abstract and theoretical"; "too much effort is needed to implement them"; "the immediate benefit is not perceived", and stresses that methods need to fit the needs of designers and their working practices. Yet, many companies see design methods as an integral part of their ways of working (López-Mesa and Bylund, 2011), even though many engineers are often not even aware that they are using specific methods (Gericke et al. 2016), as methods are often only used as long as they seem useful. Each company uses a slightly different set of methods, as methods come in with recent graduates from different universities or past employers of new recruits. Practitioners also actively look for methods when they encounter a problem or crisis (Otto 2016); however the direct introduction of methods into industry has remained a challenge (e.g. Jagtap et al. 2014).

Many methods are too complicated to use in their entirety, so that companies pick the elements that they find useful. Methods also contain many hidden assumptions, for example about how a particular element is tackled, that people from different disciplines or backgrounds might not be familiar with. Each discipline and company has its own traditions. Mechanical engineering or product development as well as software development have strong methodological foundations. Science and mathematics have strong traditions of repeatable procedures, for example how to carry out a test or calculate a values, but less so for how to carry out chunks of processes.

This paper reflects on the lived practice of two of the authors, who are dealing with an interdisciplinary work context on a day-to-day basis as managers of engineering teams with combinations of experts from different disciplines. They have selected specific examples to illustrates some of the complexity of facilitating work across different disciplines that they observe in their daily work. Their own background is in product development and mechanical engineering. The case studies are a summary of their experiences and are not independently validated.

2 Background

Many of the challenges of our time, as for example articulated in the United Nations sustainable development goals, require contributions from different disciplines, that transcend their boundaries (Moallemi et al. 2020), where communities and policy makers come together with creators of technical solutions. These challenges also require the development of new and substantially different products, which in turn bring people together from different disciplines. This requires a greater integration of disciplines in terms of their knowledge, but also their ways of working. This section starts by providing some background on the nature of collaboration across disciplines and then summaries current debates on methods in engineering.

2.1 Interdisciplinary Design

Work across different disciplines or in collaboration between disciplines is various described as multidisciplinary, interdisciplinary or transdisciplinary, however there are subtle differences between the terms. Building on Nicolescu (2008) Eckert and Stacey (2021) define "

- Multidisciplinary activity is the application of the concepts and methods of different disciplines to a common problem, in a more-or-less coordinated way.
- Interdisciplinary activity is the application of the concepts and methods of one discipline to a problem posed and framed by another discipline: a tighter coupling of concepts and paradigms than is suggested by the term 'multidisciplinary'.

- Transdisciplinary activity blurs and transcends traditional disciplinary boundaries to achieve a more fluid synthesis, or a set of problem formulations, concepts and methods that defy conventional disciplinary classification".

The economist Manfred Max-Neef (2005) sees multidisciplinary activity occurring when different disciplines are working on the same area without coordination, while he sees interdisciplinary work as a coordinated of different activities. He argues that a transdisciplinary approach would combine aspects and disciplines operating on each level. Interestingly he positions engineering on the purpose level, but design on the normative level. A strong transdisciplinary position includes for Max-Neef the coexistence of multiple viewpoints and principles, where different levels of reality and contradictions can exist and complexity principles apply.

According to Nicolescu (2006) interdisciplinarity is concerned with the transfer of methods from one discipline to another, but remains within the boundaries of those disciplines, whereas transdisciplinarity is concerned with a unity of knowledge that goes between, across and beyond disciplines. Russell et al. (2008) advocate that universities focus on developing the capacity for transdisciplinarity rather than the knowledge "products" themselves. This can be achieved by people being training in multiple disciplines, i.e. combining the disciplines in one brain, which however runs the risk that students are overwhelmed with content or end up without a clear disciplinary identify. Alternatively, university can take an application focus, where students from different disciplines come together in integrated teams or students from one discipline take a particular disciplinary perspective.

There is a growing body literature on interdisciplinary and transdisciplinarity in engineering (Lattanzio et al. 2020) in particularly around sustainability and resilience, where the understanding that engineering problems need to be tackled in a holistic way incorporating both social science, natural science and engineering is most advanced. Wognum et al. (2018) state that transdisciplinary engineering research "aims at integration between and alignment of knowledge from different disciplines, as well as theoretical and methodological transformation of each discipline throughout the process of the research". While universities increasingly make an effort to offer transdisciplinary design courses (Liu and Tran 2022).

The design of large-scale complex engineered systems requires an interdisciplinary approach, where "the practices and conventions of each discipline are interactively blended" (McGowan et al. 2013). In a survey of 62 experts at Nasa the "vast majority" of participants find interdisciplinary working leading to "confusion", "more coordination" and "communication barriers", leaving them with a sense of disorderliness. They accordingly say saw "culture", "language barriers" and "misalignment of objectives" as the major barriers but saw of "broader understanding" and "emerging thoughts" as main benefits (McGowan et al. 2013). In this case study the issues had also been aggravated by geographical distance between disciplinary teams, but could be alleviated by intermediaries who can broker between different disciplinary groups. Engineering educators are aware of some of these challenges. Efforts in teaching interdisciplinary learning often focus promoting teamwork and collaboration, for example for shared student products, and fostering social awareness and entrepreneurial skills (Van den Beemt et al. 2020).

Interdisciplinary degrees are beginning to be offered for new emerging topics, such as robotics or biomedical engineering.

2.2 What is a Method?

While we might intuitively know what a method is, the concept is interpreted differently in different communities. Pahl and Beitz (Pahl et al. 2007, German version p. 784) define a method as a *"systematic procedure with the intention to reach a specific goal"* (in German "planmäßiges Vorgehen zum Erreichen eines bestimmten Ziels"), while Cross (2008) sees all observable ways of working, which can be "procedures, techniques, aids, or 'tools' for designing" as design methods. Daalhuizen et al. (2019) see a method rather more narrowly as a formalised representation of a design activity, which functions as a mental tool to support designers in achieving a goal, in relation to the circumstances and resources available".

Academia often develops methods in isolation, but engineers in industry need to deploy individual methods in the context of other methods, which sometimes come from different communities. It is therefore important to think of methods as forming parts of an ecosystem of methods, which operate with similar vocabulary and assumptions and provide each other's inputs and use each other's output (Gericke et al. 2020). Methods not only need to be part of ecosystems of methods that operate in organisation or across interconnected supply chains, but also can be part of methodologies for addressing particular problems. In software design a methodology is typically thought of an overarching approach for creating an artifact that specifies activities, methods, how to sequence them and what information they generate. This meaning is increasingly used across different fields of engineering.

According to Gericke et al. (2017, 2022) a method contains multiple elements (see Fig. 1):

- a *core idea:* the basic activity or transformation that can be achieved through the method the core idea includes the set of concepts that people need to think with when using the method and includes.
- a *procedure*, a sequence of the activities that need to be carried out in order to achieve a result.
- Specifications of *representations* of particular types of information should be used with the method.
- In some cases, suggestions for *tools that might be useful in carrying out parts of methods* are included; in engineering these are often software systems that automate parts of the activity or manage information.
- A description of the *intended use* of the method

Daalhuizen and Cash (2021) put forward a *Method Content Theory*, which divides the content of a method into *method goal, method procedure, method rationale, method framing* and *method mindset*. They draw on Andreasen (2003) for the idea of method mindset, which they define as "the set of described values, principles, underlying beliefs, and logic that inform a method and its use".

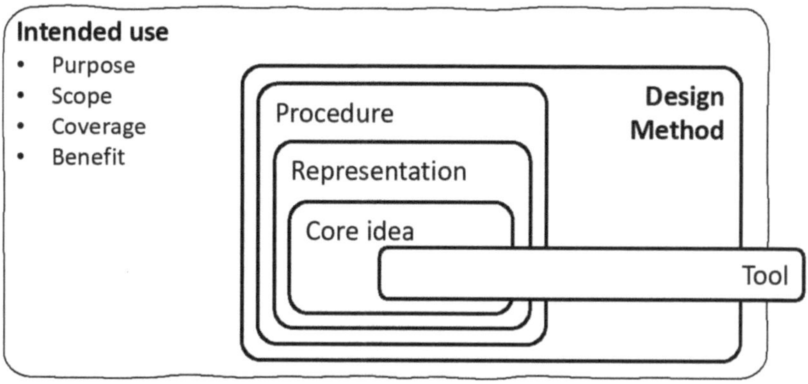

Fig. 1. Elements of a methods from Gericke et al. (2017)

2.3 Methods with Different Scope and Scale

The boundary between method and methodology is fluid, as methods occur at different scales from small scale timebound activities, such a brainstorming, to highly involved multi-step activities like QFD. While some methods are applicable in many contexts, others are highly specific to a particular work context or phase of the design process. Many industry sectors have specific procedures that they must follow, either because they are mandated by regulation or demanded by customers.

In practical applications the users of methods rarely follow all the steps of a method, but pick what is useful to them. For example, few companies go through all the steps of QFD, but many make use of the house of quality idea, which relates requirements to elements of the product architecture.

Engineers in company are rarely able to point to the specific method that they are using, but are strongly influenced by the methods that they have been taught at university or picked up from their colleagues (Gericke et al. 2016). The tacit nature of the method understanding of many engineers makes it difficult to share method in teams, because rather than being able to point to a method, that a colleague could read up on or be instructed in, they assume that how they have learned learned to do something is the way to do it. Explicit methods can be scrutinised and improved, which is much harder with internalised and to some extend tacit method knowledge; and therefore can give rise to thinking that others are doing things "wrong", because they approach problems with a different mindset from different backgrounds.

This points to an important distinction between a mindset where methods are expected to be followed step by step, and a mindset where a method provides a systematic way of approaching a problem, but is open to be adapted and extended according to the needs or disciplinary background of the team members. The narrow interpretation of methods, as a set of steps to be followed, makes engineers reluctant to use methods when they are confronted with what they perceive as a standard problem, even though following a method might reduce sources of errors.

3 Case Studies

Interdisciplinary working is required in many different work contexts, as products that have once been developed by a single discipline are increasingly becoming more interdisciplinary (Isaksson and Eckert 2020). The following case studies come from different sectors and have different challenges in integration of different disciplines. The production machine builder was founded by a mechanical engineer in 1923 and up to the 1980s was developing building and selling a mainly mechanical and electrical product, which gradually introduced more computer control. Now the products incorporate a sophisticated computerised control on multiple levels from controlling each sensor in the machine to data storage for whole production plants in the cloud. Data is collected from the machines during use to enable remote monitoring and improvements. This evolution brought a large number of software engineers and specialised computer scientists into the organisation. The coating manufacturer specializes on developing and producing technical coatings for the fastener industry. The coating is based on chemical development, but the performance of a coating is required in purely mechanical products or systems. This requires close collaboration between chemists and mechanical engineers.

3.1 Case Study 1: Complex Production Machine

Over the past decades the complexity of the product architecture has been growing and a broad range of different business models has been emerging. This makes a consistent and efficient collaboration between disciplines crucial to stay globally competitive, especially for manufacturers of complex technical products. Shorter release times, with new functionally and new features, leave the company little time for change processes. The following three examples illustrate some of the challenges in disciplinary integration the company faces.

- **Classical machine development** still follows the classical development process as taught in mechanical design in university. However, in the last years the process was restructured to provide half-yearly releases to incorporate changes to the whole machine portfolio arising from the company's platform strategy. This release process is currently dominated by the release dates for physical components. Software and electrical engineering follow this process and its release cycles, but lose time through it, because shorter releases with single features or smaller software updates are hard to carry out. The different professional groups have a different understanding of the functional decomposition of a machine and negotiate solutions locally.
- **Digital twins and virtual testing in the development process**: The strong emphasis on physical development also limits effective digitalisation and use of virtual tools throughout the entire development process. Building virtual twins still requires high effort, so that the digital twins are often only ready once the physical prototypes are already built, so that virtual simulation and testing cannot be used to maximal benefit. This time lag is caused by virtual methods not being systematically used in design teams, but deployed differently depending on expert know-how. These competences are not shared systematically within the organisation. This is particularly critical as the understanding of validation differs between classical machine development,

where a whole machine is tested at the end of a release, and continuous testing of new software components as they are being developed. The value of a new software version is generated out of hundreds of individual software components. However, whether the software works in the context of a newly designed machine can only by established during final testing when the software and the hardware are tested together.

- **Integration of data-based services:** Over the last years data-based products have extended the traditional product scope and have become an important part of the business. Developing data products imposes even more challenges on collaboration as it brings in data scientists and mathematicians. The understanding of the data lifecycle as well as the methods and infrastructure to build data products are fundamentally different to classical engineering methods and processes. The team members often come with totally different backgrounds and requirements on collaboration. However, data products make it necessary that all areas work together: data of the required quality has to be provided by the classical machine components, in particular sensors; it has to be processed to be further analysed and used for other purposes. Data products have a fundamentally different life cycle, as they are run continuously and need sufficient care and infrastructure.

3.2 Case Study 2: Specialised Coatings

The company develops and manufactures coatings for fasteners, which fulfil particular functions such as friction control, sealing or locking. Locking coatings for screws are a particularly important product for the company. These functions are realised with thread coatings with encapsulated reaction partners. Using microencapsulation technology droplets of the multicomponent system are surrounded by a thin polymer wall. In locking coatings, the reaction partners are incorporated into a coating similar to a paint, which once applied and dried results in a dry non-tacky coating that is permanently attached to the threaded part. The coating remains inactive until the threaded part is installed. The pressure and/or shear stress during the assembly process ruptures the microcapsules in the coating. The liquid components are released and the mixture cures in conjunction with the binder system.

One of the authors, who is the team leader, recalls: "When I look back to my time at university I do remember very well, that we (mechanical engineering students) did not like the other disciplines we had to take exams in: electrical engineering, chemistry, … We did see ourselves as mechanical engineers. We were not aware, and no one told us, that in industrial R&D, most of the projects involve different disciplines. I particularly hated chemistry – and I ended up in a company where chemistry and mechanics could not be more connected." This attitude is similar to that of his younger mechanical engineering colleagues. The R&D team includes mechanical engineers as well as chemists. Due to their education chemists are familiar with knowledge and methods that are very different, and have a very different approach to addressing problems. As scientists their understanding of methods is of techniques or processes for creating a particular kind of result. Traditionally scientific methods were mostly conceived as strict rules or procedures; even though this notion has been eroded in theory (Hoyningen-Huene 2008), there the expectation of methods yielding specific results remains. For the scientists the looser concept of an engineering method is quite alien. While they are used to standardised repeatable procedures, they typically don't use open opened methods for example to generate ideas or use methods to structure or plan their processes. To develop new products in mixed discipline teams a basic understanding of the other disciplines and their ways of working is a basic requirement.

4 The Use of Methods Across Domains

The case studies illustrates that there is a divergent understanding of what a method is for in the different disciplinary groups, without the engineers necessarily being aware that this is the case. However, it also points to deeper differences in people's understanding of what the product development process is, and what its purpose is. This section analyses some of these points of divergence. These also exist within a single discipline, but are aggravated by the different disciplines the team members come from.

4.1 Single Tasks Involving Multiple Disciplines

Modern product development has become extremely integrated, so that seemingly straightforward tasks, such as taking a sensor reading and responding to it, involves the contribution of multiple disciplines as well as subdisciplines in one discipline. For example, in case study 1, the company provides data from sensor signals to their customers, which shows whether the machines are running or not. To do so it has to go through a process involving many different specialties:

Table 1. Disciplines Collaborating on a Simple Task

Step	Task	Specialty	Discipline
1	Obtaining the data from the machine	sensor technology, sps-software,	electrical engineering
2	Sending the data across the right pipelines	Data pipelines architecture	Electrical engineering
3	Preparing the data for analysis	data cleaning	Data or computer science
4	Carrying out data analysis	diagnostics	Mechanical or electrical engineering
5	Display on a suitable front end	UX design	Computer science

As Table 1 illustrates, this seemingly simple process involves team members with different kinds of engineering or computing degrees. Even if they have the same academic background, there are huge differences in how a software engineer or a data analyst approaches a problem. This poses a huge communication challenge, as team members make assumptions about the knowledge and working practices of their colleagues and the terminology with which this is expressed.

4.2 Different Backgrounds and Contexts

Engineering companies bring together engineers and other experts from a very wide range of backgrounds which influence how people think about their activities.

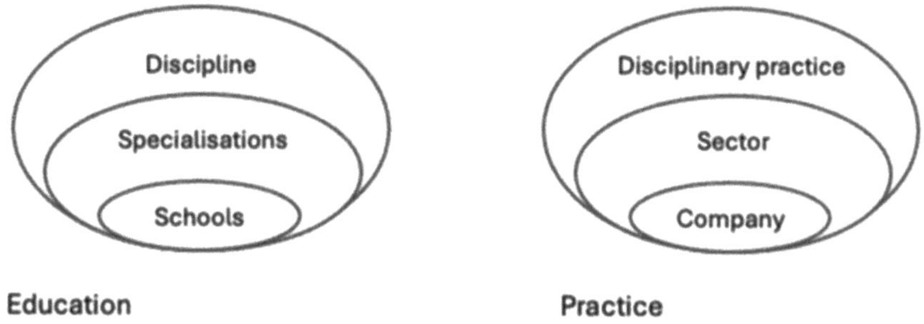

Fig. 2. Nested Identities of Interdisciplinary Teams

Figure 2 illustrates the individual team members have nested identities. Even within one discipline, the team members are trained at particular universities, who teach many different tools and methods within one degree. During their degrees or in their professional practice, the team member will have specialized to a certain extent, for example in product development or stress engineering. Some people also specialize in a particular product class, such as aerospace engineering or cybersecurity. However, their work context also influences them and give them a nested identity. Each company has its own way of working, which is partly determined by the characteristics of the products the team members come into contact with, but there are noticeable differences between competing companies. The industry sector in which the company operates also affects the way a company works, for example automotive companies work in different ways to aerospace companies or production machine companies. The influence of these sectors are embedded in the disciplinary practice of the company, often going back to the origins of the company. For example, the production tool company originally made mechanical products and is still dominated by mechanical engineers in the senior and middle management, even though it is trying to be more inclusive of different disciplines. Figure 2 give a structure to the object worlds of individual engineers and point to the places of overlap between different members of team. While object worlds are personal, the overlap also gives the opportunity to translate between inhabitants of object worlds.

Larger organisations, like the tool builder in case study I, often have liminal people, who can understand the object worlds of different groups. These can be line managers, who have made it their business to understand the perspectives of different groups, or people who have been trained in working across these disciplines, for example through a joint degree. Smaller companies, like the coating manufacturers, who work across two disciplines, can try to bring the entire team to a shared understanding and negotiate the use of common methods. In the coating manufacturer, the chemists have to understand the required properties of the coating during the coating process, but also the function of the coating in the system the fastener is being used for. When the chemists understand the mechanics during the installation process, they can optimize the adhesion or cohesion much better. Conversely, if the engineers do not understand the chemistry, they cannot give valuable feedback.

4.3 Lack of Shared Vocabulary and Representations

These different backgrounds define different vocabularies and different processes that people are familiar with. For examples, there is a lively debate on what is meant by the term 'function' within the engineering community (e.g. Eckert 2013); however, there is an even greater difference between disciplines (see Eisenbart et al. 2012). Mechanical engineers typically think of a function in terms of means of fulfilling a particular goal, whereas mathematicians think of function as a deterministic transformation of inputs into outputs, whereas computer scientists think of functions as sequences of instructions that compute particular results. Therefore, a function for a computer scientist or a mathematician is inherently repeatable and once implemented it can be used in multiple places. Whereas a mechanical engineer assumes that functions are phrased in a solution neutral way and can be realized in many different ways; therefore they see a need to revisit the same function in different products. This different understanding of function leads to frustration in case study 1. The computer scientists wonder why the mechanical engineers keep designing new components for standard functions, and why existing components needs to be tested in new contexts. While the mechanical engineers acknowledge that sometimes a greater standardization might be possible, they focus on overall system performance and therefore redesign components to improve system parameters, like speed or robustness.

Each academic discipline has its own representations that everybody is familiar with, but which are not necessarily accessible to people from different fields. For example, mechanical engineers use CAD models as one of their primary representations, which give elements of a system a specific location in the context of the wider product, while computer scientist frequently use UML diagrams, which give a logical, but not a spatial connection. This is particularly an issue in the interaction between the mechanical engineers and the chemists, where the mechanical engineers have to explain the mechanical connections of specific products to chemists who can read technical drawings or understand CAD models.

4.4 Layers of Shared Understanding

With the different object worlds, experiences and disciplinary perspectives of different engineers also come different understandings of the problems they are trying to address. Many authors have argued that a shared understanding is required to carry out design projects successfully (Hill et al. 2001; Eris et al. 2014)). Personal experiences of individuals can influence the collective narrative of a design situation (Lloyd 2000). Asking questions and getting feedback is major antecedent for shared understanding (Cash et al. 2017). Several constituent elements of shared understanding have been identified in the literature including a shared vision (Chiu et al. 2006), a shared solution understanding (Preston et al. 2006) and a shared understanding of the roles, responsibilities and concerns of the team members (Badke-Schaub et al. 2007).

For example, when applying the chemical coating to the screw the viscosity of the coating is critical, as it is when they screw is deployed. This depends both on the detailed geometry of screw and the chemical properties of the coating and needs to be developed in together. In addition, it is important to understand exactly what is happening to the coating as the screw is screwed in and during the manufacturing process. The success of the product depends on all having at least a superficial understanding and some understanding this interaction in great detail.

Our case studies revealed that there was a lack of shared understanding between the different professional groups on multiple levels. These are interconnected, but involve distinctly different perspectives:

1. **Understanding of the purpose of the product or activity**: While there might be a consensus about the core functionality of the product, the understanding of the importance of particular features of the product can vary enormously. For example, for designing suitable screw coatings it is absolutely critical that the chemists understand the details of the mechanical processes in the screw and the variability of the use context. Design engineers understand different use cases and worst-case scenarios, whereas a scientist might focus on reaching a particular clearly defined state, such as withstanding a particular temperature range. While this can pragmatically be managed by providing clear requirement specifications, to resolve potential problems a deep understanding of the propose of the product is required.
2. **Understanding of the technical system**: this is tied up with the vocabulary in which different groups talk about the system. The understanding of the technical system involves both the scope of the system and the importance of particular features. For example, service or application engineers are thinking of the physical product together with the service process, while a materials engineer only thinks in terms of physical elements of a system. Software engineers think in terms of the functional interactions of units of the system, represented in UML or SYSML diagrams, whereas mechanical engineers are very aware of the failure modes that can arise from the physical interaction of components, e.g. through vibration under extreme operating conditions.
3. **Understanding of the process**: The sequence of tasks and the importance of specific tasks also depends on domain practices. This is illustrated by the role and place of testing in product development processes, as illustrated in case study I. Software engineers test their software as they go along, and assume that if elements work well, the interactions in the whole system work in a predictable way. While mechanical engineers also test through the product development process and use modelling and simulation (see Tahera et al. 2019), they depend heavily on physical testing of the integrated product and therefore are very concerned with meeting product deadlines, that enable them to meet prebooked slots on testing rigs. Another source of difference between disciplines is how they handle their change processes and their release management. For software frequent releases of updates to running systems are common and both designers and customers are used to them. However, the integrity of physical systems can be compromised by changes, so that the products need to be retested. This is one reason why mechanical engineers aim to bundle changes and have bigger releases at larger intervals.

5 Introducing Methods Across Disciplines

The understanding and use of engineering design methods needs to be understood in the context of the lack of a shared understanding of processes.

The methods used in organizations in a formal way are often determined by the management of an organization, for example through the purchase of supporting tool suites which are rolled out across the entire organization, or by line managers based on their experiences. As the management of companies is still recruited from traditional disciplines, for example mechanical engineering in case study I, these disciplines have a long-lasting effect on the choice of tools and methods that an organization uses officially. When new disciplines enter an organization, it can take a while before representatives join management. For example, while industry currently hires a lot of data scientists, few will have become managers. Unless they have experience of working with methods of other fields, they might not see the need behind the method preferences of their colleagues.

While there is a strong disciplinary bias for methods, there is also fads for methods across organizations. For example, Six Sigma, which had come out of quality control for mass manufacturing, was widely used in the early 2000s. Currently there is a huge fashion for agile methods (e.g. Heimicke et al. 2021), which has come out software development of interactive systems where frequent software updates and quick user feedback are very helpful. In this case methods are either pushed by young enthusiasts or introduced by managerial dictum, so that teams have to figure out how they can implement the approach.

The introduction of methods can be a challenge in practice on multiple levels:

- **Understanding of working with methods in general:** In some fields, in particular in science, there is little understanding what a method is in engineering. Scientists are familiar with repeatable procedures, for example in experiment set up, that need to be followed in detail to assure success. However, engineering methods are often looser; they require interpretation and adaptation to a specific context, as they are a form of abstracted knowledge. They produce guidance about the sequences and dependencies of steps and indicate the type of result that can be generated rather assuring a specific result.
- **Understanding the mindset behind particular methods:** Even when people are familiar with methods in general, the vocabulary and assumptions of the methods that come out of domain practice might be difficult to understand.
- **Understanding the details of a specific method:** Methods descriptions are often incomplete as they assume a general understanding of the field or leave out particular steps that some participants might be not familiar with.

Therefore, it cannot be taken for granted that team members with different background can pick up methods, when they are given method descriptions or oral explanations. Method introduction requires active and sensitive management. Team members need to be persuaded. Methods need to be explained to them in a way they can relate to and their concerns need to be listened to. This is often not the case in practice and teams muddle through by incorporating chunks of methods that seem useful in their own way of working. In particular, teams need examples of successful use of methods, to be persuaded that the methods are in fact useful and that it is worth sticking with different elements of a method.

The successful use of methods depends on an understanding of the underlying process. The coating company started by defining a systematic product development process, that defined the basic steps of the process and provided a context for specific methods. The production tool company has had processes for a very long time, but needed to understand what the understanding of the technical systems of different professional groups was, to be able to tailor the methods and the explanation of methods to the needs to each group.

6 Implications for the Development of Methods

This raises questions for the methods that industry has to work with. There is a lively debate about the challenge of method introduction; however there are a few points that the academic community needs to consider:

- Are methods too complicated for industry? The methods need to be sufficiently simple and intuitive to pick otherwise the teams and their managers need to put considerable effort into explaining methods and persuading people to use them. This requires good explanation in the method descriptions. This is particularly important for small companies, who might not have experts who can dedicate themselves to method introduction. Big companies typically have method groups; however, they have to "sell" methods internally and also need support in doing so.
- Are we clear about the background knowledge assumed in the use of a method? In interdisciplinary teams it cannot be assumed that everybody has even the same basic knowledge. Method descriptions usually don't come with pointers on how to fill gaps.
- Do we know when a method has been successful? Companies usually use method as long as they find them useful, and often abandon them when they have got what they were looking for, when they get stuck, or when they see no benefit. Methods are rarely followed from beginning to end. Even if methods are completely applied, they many show a path to a solution rather than guarantee a particularly successful one.
- Should methods be developed by interdisciplinary teams? As long as methods are still developed by researchers or industry teams from a single disciplinary perspective, it is difficult for others to fit in.

7 Conclusion

The paper has argued that the introduction of methods in interdisciplinary organizations can be challenging because of divergent understanding of methods and too little experience with methods in practice. Some of these issues can be addressed by careful and insightful management and persuasion of individuals. However, the methods themselves must also become more accessible to all the potential users. This implies that methods need to be tested with users from different backgrounds. Maybe the most important point is that methods need to be kept as simple as possible even if this requires more effort and iteration in the development of methods.

However, universities can also support the interdisciplinary use of methods. While some universities are offering cross degree or even cross faculty projects or put teams together for particular challenges, the interdisciplinary training is still far away from everyday industrial practice; and students are not explicitly taught about the perspectives and assumptions of the disciplines they need to work with. The collaboration across disciplines has become such an integral part of modern work practice that students need to be trained for it.

References

Andreasen, M.M.: Improving design methods' usability by a mindset approach. Hum. Behav. Des. Individ. Teams Tools, 209–218 (2003)

Bucciarelli, L.L.: Between thought and object in engineering design. Des. Stud. **23**(3), 219–231 (2002)

Badke-Schaub, P., Lauche, K., Neumann, A., Ahmed, S.: Task–team–process: assessment and analysis of the development of shared representations in an engineering team. In: DTRS 7 Design Meeting Protocols Workshop Proceedings, London, pp. 97–109 (2007)

Cash, P., Dekoninck, E.A., Ahmed-Kristensen, S.: Supporting the development of shared understanding in distributed design teams. J. Eng. Des. **28**(3), 147–170 (2017)

Chiu, C.M., Hsu, M.H., Wang, E.T.: Understanding knowledge sharing in virtual communities: an integration of social capital and social cognitive theories. Decis. Support. Syst. **42**(3), 1872–1888 (2006)

Cross, N.G.: The design process. In: Engineering Design Methods: Strategies for Product Design, 4th edn., p. 230. Wiley, Chichester (2008)

Daalhuizen, J., Cash, P.: Method content theory: towards a new understanding of methods in design. Des. Stud. **75**, 101018 (2021)

Daalhuizen, J., Timmer, R., Van Der Welie, M., Gardien, P.: An architecture of design doing: a framework for capturing the ever-evolving practice of design to drive organizational learning. Int. J. Des. **13**(1), 37–52 (2019)

Eckert, C.M.: That which is not form: the practical challenges in using functional concepts in design. AI EDAM **27**(3), 217–231 (2013)

Eckert, C., Stacey, M.: Overconstrained and underconstrained creativity: changing the rhetoric to negotiate the boundaries of design. In: Blessing, L., Qureshi, A.J., Gericke, K. (eds.) The Future of Transdisciplinary Design, pp. 207–219. Springer, Cham (2021). https://doi.org/10.1007/978-3-319-06382-9_16

Eisenbart, B., Blessing, L.T.M., Gericke, K.: Functional modelling perspectives across disciplines: a literature review. In: DS 70: Proceedings of DESIGN 2012, the 12th International Design Conference, Dubrovnik, Croatia, pp. 847–858 (2012)

Eris, O., Martelaro, N., Badke-Schaub, P.: A comparative analysis of multimodal communication during design sketching in co-located and distributed environments. Des. Stud. **35**(6), 559–592 (2014)

Gericke, K., Eckert, C., Stacey, M.: What do we need to say about a design method? In: Proceedings of the International Conference on Engineering Design, ICED. The Design Society (2017)

Gericke, K., et al.: Supporting designers: moving from method menagerie to method ecosystem. Des. Sci. **6**, e21 (2020)

Gericke, K., Kramer, J., Roschuni, C.: An exploratory study of the discovery and selection of design methods in practice. J. Mech. Des. **138**(10), 101–109 (2016)

Heimicke, J., Dühr, K., Krüger, M., Ng, G.L., Albers, A.: A framework for generating agile methods for product development. Procedia CIRP **100**, 786–791 (2021)

Hill, A., Song, S., Dong, A., Agogino, A.: Identifying shared understanding in design using document analysis. In: International Design Engineering Technical Conferences and Computers and Information in Engineering Conference American Society of Mechanical Engineers, vol. 80258, pp. 309–315 (2001)

Hoyningen-Huene, P.: Systematicity: the nature of science. Philosophia **36**(2), 167–180 (2008)

Isaksson, O., Eckert, C.M.: Product Development 2040. Public Report. The Design Society, Glasgow, UK (2020)

Jagtap, S., Warell, A., Hiort, V., Motte, D., Larsson, A.: Design methods and factors influencing their uptake in product development companies: a review. In: DS 77: Proceedings of the DESIGN 2014 13th International Design Conference, pp. 231–240 (2014)

Lattanzio, S., et al.: Transdisciplinarity within the academic engineering literature. Int. J. Agile Syst. Manag. **13**(2), 213–232 (2020)

Liu, Q., Tran, H.: Exploring transdisciplinarity in engineering education and practice: a review of literature and existing initiatives. Proc. Can. Eng. Educ. Assoc. (CEEA) (2022)

Lloyd, P.: Storytelling and the development of discourse in the engineering design process. Des. Stud. **21**(4), 357–373 (2000)

López-Mesa, B., Bylund, N.: A study of the use of concept selection methods from inside a company. Res. Eng. Des. **22**, 7–27 (2011)

Maier, A.M., Eckert, C.M., Clarkson, P.J.: Factors influencing communication in collaborative design. J. Eng. Des. **32**(12), 671–702 (2021)

Max-Neef, M.A.: Foundations of transdisciplinarity. Ecol. Econ. **53**(1), 5–16 (2005)

McGowan, A.M.R., Daly, S., Baker, W., Papalambros, P., Seifert, C.: A socio-technical perspective on interdisciplinary interactions during the development of complex engineered systems. Procedia Comput. Sci. **16**, 1142–1151 (2013)

Moallemi, E.A., et al.: Achieving the sustainable development goals requires transdisciplinary innovation at the local scale. One Earth **3**(3), 300–313 (2020)

Nicolescu, B.: Transdisciplinarity: past, present and future. Moving Worldviews: Reshaping sciences, policies and practices for endogenous sustainable development, ETC/COMPAS, Leusden, pp. 142–166 (2006)

Nicolescu, B. (ed.): Transdisciplinarity: Theory and Practice. Hampton Press, Cresskill (2008)

Otto, K.: Industry Adoption of Design Methods: Where is the Crisis? Cavtat, Croatia (2016)

Pahl, G., Beitz, W., Feldhusen, J., Grote, K.-H.: Konstruktionslehre Grundlagen erfolgreicher Produktentwicklung, Methoden und Anwendung, 7th edn. Springer, Berlin (2007)

Preston, D.S., Karahanna, E., Rowe, F.: Development of shared understanding between the chief information officer and top management team in US and French organizations: a cross-cultural comparison. IEEE Trans. Eng. Manag. **53**(2), 191–206 (2006)

Russell, A.W., Wickson, F., Carew, A.L.: Transdisciplinarity: context, contradictions and capacity. Futures **40**(5), 460–472 (2008)

Tahera, K., Wynn, D.C., Earl, C., Eckert, C.M.: Testing in the incremental design and development of complex products. Res. Eng. Des. **30**, 291–316 (2019)

Van den Beemt, A., et al.: Interdisciplinary engineering education: a review of vision, teaching, and support. J. Eng. Educ. **109**(3), 508–555 (2020)

Wallace, K.: Transferring design methods into practice. In: Birkhofer, H. (ed.) The Future of Design Methodology, pp. 239–248. Springer, London (2011). https://doi.org/10.1007/978-0-85729-615-3_21

Wognum, N., Bil, C., Elgh, F., Peruzzini, M., Stjepandić, J., Verhagen, W.: Transdisciplinary engineering research challenges. In: Transdisciplinary Engineering Methods for Social Innovation of Industry 4.0, pp. 753–762. IOS Press (2018)

Design Methods and Design Processes in the Digital Age

Ralf Stetter(✉)

RWU Ravensburg-Weingarten, Weingarten, Germany
Ralf.Stetter@rwu.de

Abstract. The field of engineering design is currently undergoing a profound transformation. The availability of powerful computer tools as well as increased connectivity concerning tools and products have not only reshaped the tools and methods engineers use, but also have considerably altered the processes through which innovative products are conceived and developed. This paper reflects this trend and is consequently a collection of prominent changes as well as an analysis of the drivers and consequences of these changes and a deduction of implications concerning future design processes and design methods. One main driver for change is the presence of some form of digital design process management. Today, in most engineering companies, a digital design process management is implemented, which prescribes design synthesis results such as required product models with their respective level of maturity and also prescribes analysis results such as required simulations. Additionally, milestones, mechanisms for version, variant and configuration management as well as further aspects of the product life cycle management are included in this digital design process management. Through the application of the digital design process management considerable advantages in terms of process transparency and time to market can be achieved and deficiencies such as redundant data storage or contradictions can be avoided, but the processes of design engineers are considerably altered. It is desirable that the needs of these design engineers are considered in the synthesis of the digital design process management. The second main driver for the current changes in the design processes is an increase of the possibilities for fast prototyping. This covers both in the virtual domain with simplified finite element analyses within the design system (usually CAD system) and the physical domain with fast and easy possibilities for 3-D-printing. As a third prominent driver for the current changes in the design processes, an altered communication behaviour can be identified. An important influence was the COVID pandemic, but also before that online meetings were frequent in internationally operating companies. Meanwhile, designers communicate frequently using digital collaboration platforms and virtual as well as hybrid meetings seem to have become the preferred mode of collaboration. This paper scrutinizes the manifold consequences of these three drivers and explains how these should be considered in future design processes.

K. Gericke et al. (Eds.): DSC 2022/2023, PTT, pp. 36–52, 2026.
https://doi.org/10.1007/978-3-032-16278-6_4

1 Introduction

Has the daily work of design engineers in industrial companies changed in the last years? Yes, it has – and maybe more than in the decades before. In recent years, new digital technologies have continuously emerged, constantly posing new challenges for companies, their employees and society; the five most representative technologies are the Internet of Things (IoT), cloud computing, big data, social media and mobile technology (Zhu et al. 2021). Another current development is the application of artificial intelligence (AI). AI enables automating functions that are vastly associated with the human mind, such as sensing, perceiving, interacting, problem-solving, learning or decision-making (Benbya et al. 2021). This contribution collects prominent changes, analyses the drivers and consequences of these changes and deducts implications for future design processes and design methods. One main driver for change is the presence of a digital design process management which usually includes working steps in synthesis (e.g. required product models with a respective level of maturity) and analysis (e.g. required simulations), milestones, mechanisms for version, variant and configuration management as well as further aspects of the product life cycle management. Another driver is an increase of the possibilities for fast prototyping both in the virtual domain (fast finite element analyses (FEA) carried out by the designer himself/herself) and the physical domain (radically improved possibilities for 3-D-printing). Yet another driver is the altered communication behaviour – started earlier and intensified through the COVID pandemic, designers communicate frequently using digital collaboration platforms. Virtual meetings, which used to be the exception, seem to have become the preferred mode of collaboration. The consequences are manifold and should be taken into account in future design processes.

Since several decades, digital tools are widely used in engineering design. Nearly all products are modelled in 3D computer aided design (CAD) systems and data is stored in specialized data management systems such as product data management (PDM) systems. Still, in the last years an intensified digitalization can be observed. Several researchers highlight that trends such as intensified competition, increasing system complexity and digitalization currently lead to enormous changes concerning the working environment, working style, engineering approaches and required competencies (Gräßler et al. 2021). This contribution will highlight some of these changes and will discuss consequences. For design research a conscious consideration of these changes is very important because the working environment and style of design engineers is changing considerably and methods, tools and processes have to be adopted to the changing situation. One of the cornerstones of digital processes is the capability to store a large share of the available data and information in a digital format. One may ask, if the changes, which are caused by the current digitalization efforts, lead to a design process which captures all essential ideas, issues, background information and process management elements digitally? From these considerations, a central research question can be derived:

Which changes have influenced the working environment and style of design engineers in the last years, what are the driver of this changes, what are the consequences of these changes and which implications for future design processes and design methods can be deducted?

Consequently, the chapter is structured as follows. Section 2 explains current changes, while Sect. 3 analyses the drivers of these changes and Sect. 4 their consequences. The deduction of implications for future design processes and design methods is described in the final Sect. 5.

2 Current Changes in Product Development Processes

This section seeks to explore current changes in product development processes in the areas of model based systems engineering (MBSE), new design paradigms, digitalization and digital integration, artificial intelligence (AI) as well as additional domains, requirements and approaches.

2.1 Model Based Systems Engineering

Today, in most industrial product development processes, a kind of model based systems engineering (MBSE) is applied in order to enable, amongst others, structured and synchronized processes. The prominent use of models in MBSE provides a common language for communication and collaboration throughout the product development process. In general, the application of MBSE can lead to improved transparency and can facilitate a systematic model management in increasingly complex processes. The combination of systems engineering principles with the transition from a document-centric to a model-based approach can allow to meet current industrial demands (Gräßler et al. 2021). The general structure of MBSE follows the well-known V-model (Fig. 1).

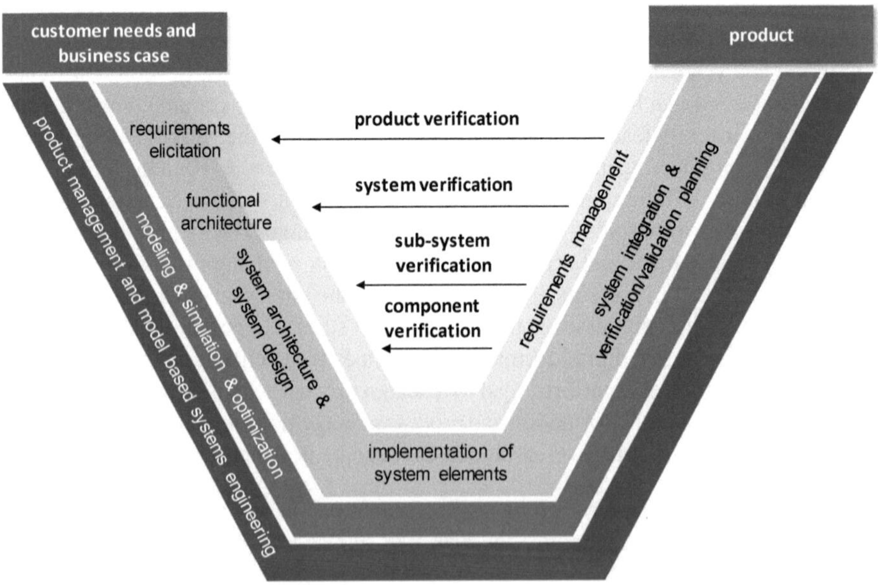

Fig. 1. The general structure of MBSE

The underlying structure of MBSE is the well-known V-model with its emphasis on domain spanning concept phase, systems integration and system verification. The whole process is supported by modelling, simulation and optimization. An important rationale for using modeling and simulation in the scope of MBSE is their potential to reflect multifaceted perspectives and interests; through transparency and actionability, simulation models can create new information and help engineers and managers make complex design choices (Lakemond et al. 2021). Another important support layer is realized by product management and the management aspects of MBSE. MBSE can be applied from early stages such as initial requirement management up to concrete physical engineering and the purpose is mainly to improve the information flow between the design engineers (Salehi et al. 2018). One would expect mainly positive effects of a MBSE introduction. However, the processes of design engineers change considerably, because new roles appear (compare Gräßler et al. 2021) and because additional models e.g. on the functional level have to be developed. Additionally, certain systems engineering artefacts, such as life cycle models, are to be generated and a responsibility assignment for the artefacts as well as an assignment of the systems engineering roles need to be developed (compare Gräßler et al. 2021). In the resulting complex, interconnected process models, it may be a challenge for single design engineers to keep an overview and to be aware about their current process contribution.

2.2 New Design Paradigms

One of the drivers of digitalization is frequently an increasing variability of the products. More and more product characteristics have to be adapted to individual needs of consumers leading to large product families. This leads to a need for a paradigm shift – from the conventional computer aided design (CAD) of a single product to the programming of a whole product family. An engineering framework, which allows the programming of product families may be based on graph-based languages. The central application aspects of graph-based design languages are depicted in Fig. 2 (published under CC license in Stetter 2020a).

The basis of the design process with graph-based design languages are ontologies which capture the knowledge in certain engineering field (compare Arnold & Rudolph 2012). A project independent data model is generated using vocabulary (e.g. system components, structure (i.e. the product architecture)) and sequences of rules. A project-specific data model is created by a so-called design compiler leading to a design graph including all information which concerns the product. This information can then be transferred to conventional engineering tools.

It is important to note that the tools, but also the mental discourse of the design engineers, is quite different from a conventional design process. The central "mode" of expressing the design intention is not a CAD geometry but entities described in a unified modelling language (UML) class diagram (other approaches use the system modelling language (SysML compare Weilkiens 2011) for instance the approaches reported by Albers & Zingel (2013) or Mhenni et al. (2014). In the initial phase, design engineers do not describe concrete product characteristics, but possible characteristics and their interconnection. The combination of models of the system of objectives following MBSE approaches with semantic descriptions in the form of a knowledge graph are not only

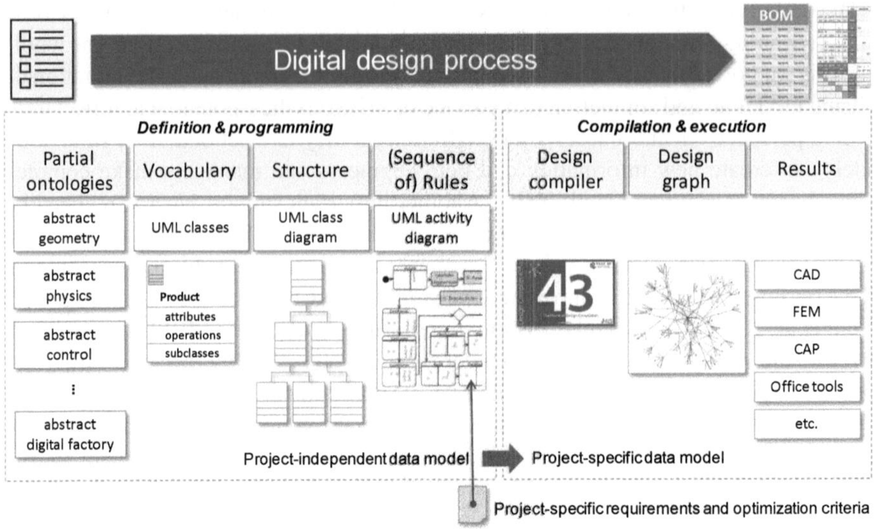

Fig. 2. Central application aspects of graph-based design languages

appropriate for supporting the development of a single product generation but also for assisting to continuously generate and process knowledge in a reusable form (Mandel et al. 2023). In addition to the design of the respective product and systems, design processes need to provide a basis for an entire product realization spectrum from marketing and customer concerns to the production processes, further to the downstream logistics and service issues, and further closing the loop with a circular product lifecycle involving many recycling, remanufacturing, repurposing and recovery issues (Jiao et al. 2022). Needless to say that the initial modelling effort may be larger and that a deeper understanding of the interconnection of certain parameters and topologies is required. To conclude, the intention to create a design "program" for a whole product family instead of firstly modelling a single product concretization may cause a paradigm shift for design engineer and may require a quite different procedure style.

2.3 Digitalization and Digital Integration

Digitalization is the process of integrating digital technology and information technology systems into the entirety of a company's processes. The main intentions of digitalization are an improved operational efficiency, the avoidance of mistakes and redundant work and an efficient knowledge management. One important cornerstone of a successful digitalization of product development processes is the integration of data and information. Future product design processes as parts of digitally enabled organizations will generate a large amount of information which need to be verified, processed and managed (Berendsen & Beckett 2020; Hashmi et al. 2021). In conventional process without information integration a heterogeneous model landscape may be created, which is characterized overlapping, partially redundantly modelled information; these overlaps and redundancies can, especially when the models have to be adapted to changed system or product requirements, lead to inconsistencies, which - if recognized too late - can lead to faulty decisions (Feldmann & Vogel-Heuser 2017). It is important to note that the integration does not only concern entities in one domain, at one level of abstraction or one level of detail. In order to analyse the challenges of information integration it is sensible to analyse the different kinds of information especially in abstract domains. Figure 3 shows a class diagram of different entities in the product development system.

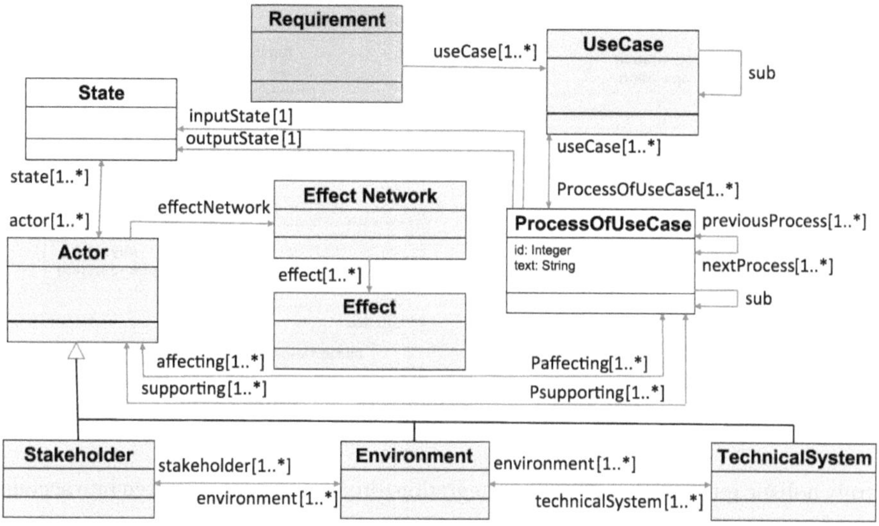

Fig. 3. Class diagram of different entities in the product development system

This representation which is based on prior work concerning function modelling (Holder et al. 2017) and modelling of abstract physics (Stetter 2020a) allows to analyse and represent the complexity of the integration endeavour. Until know several connections just exist within the brains of the design engineers, but should be ultimately integrated into a holistic digitalization concept. Figure 4 allows to deepen this discussion.

Visible in Fig. 4 are distinct levels of abstraction which conform e.g. with the Munich model of product concretization (Ponn & Lindemann 2011). Important information entities are shown, for instance at the level of product geometry. The geometrical entities are either basic geometry, features, parts or assemblies and these are generated by a sequence of certain generation operations (this is true for conventional CAD systems but also other modelling paradigms. In the requirement domain, system requirements, functional requirements, property requirements and other requirements can be distinguished. In the abstract physics domain, the functional realization of a product can be expressed via interfaces (Grauberger et al. 2020), physical phenomena (Ehrlenspiel & Meerkamm 2017) and behavior. Concerning behavior novel modular concepts are receiving increasing attention which are based on the functional mock-up interface (FMI) standard and functional mockup units (FMUs) (Larsen et al. 2016). Obviously, a holistic concept also has to include non-geometrical product information such as material and micro-geometry and process information such as description of manufacturing processes. For

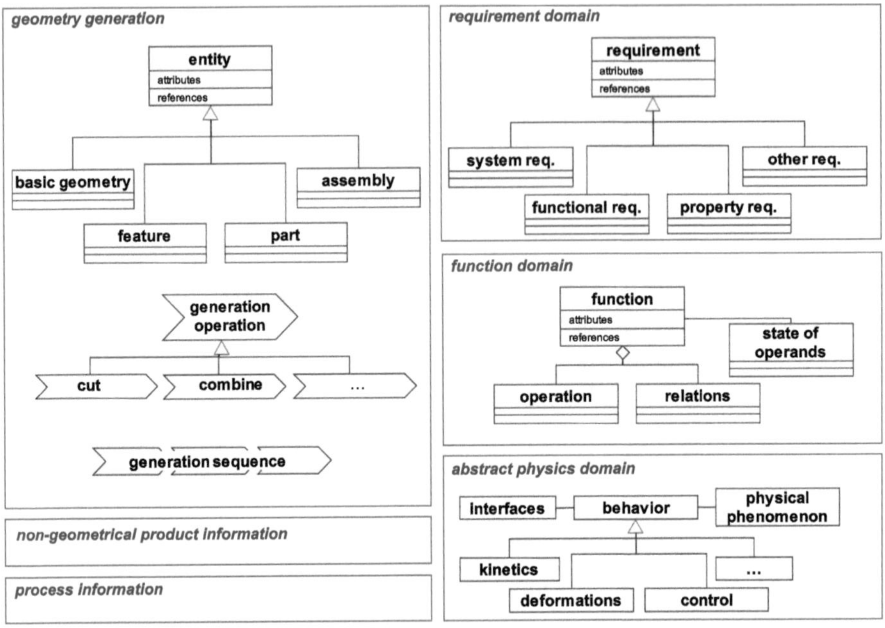

Fig. 4. Information entities for integration

a truly holistic integration, additional integration directions have to be taken into account (Fig. 5).

Figure 5 shows the integration space; three main dimensions are visible: the product concretization levels (compare Fig. 3), the scientific domains and the steps of the problem-solving-cycle, which characterize the concrete activities of the involved design engineers. It is important to note that an integration of analyses and synthesis steps is one of the main objectives of the digital integration. Further potential dimensions are the product life-cycle, variants, versions and the level of detail. Integrated models of the product and the processes necessary to generate and operate the product are desirable for several reasons such as avoiding contradictions and inconsistencies as well as the possibilities to synthesize elaborate digital twins (Trauer et al. 2021). For design engineers, additional responsibilities evolve, because the interconnections have to be modelled thus contributing to a changed demanded working profile.

2.4 Artificial Intelligence

The term "Artificial Intelligence" (AI) mainly describes machine learning applications and provides enormous innovation opportunities for companies, promising to enable the automation of cognitive tasks (Sagodi et al. 2022). Currently most applications of AI in producing companies concern the manufacturing field and are realized as isolated "proof of concept" (PoC) projects (Sagodi et al. 2022, Winkler et al. 2019). Consequently, one may ask why, in the current stage, design engineering needs to be involved with AI and machine learning. The answer can be given referring to digital twins (DT). A DT is a

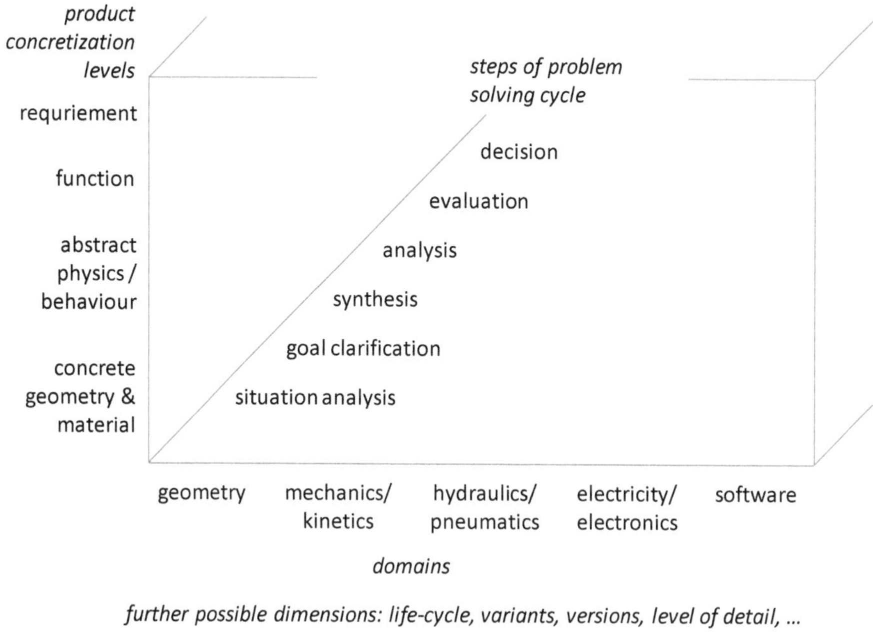

Fig. 5. Integration space

virtual representation of a real system (often an assembly or manufacturing system) that simulates certain conditions (Trauer et al. 2020) and may extrapolate it into the future. DTs are frequently discussed in the context of Industry 4.0 and may help to overcome current challenges of producing companies (Trauer et al. 2021). It is important to note that DTs cover the entire product lifecycle – also the design engineering stage (Trauer et al. 2020). This characteristic distinguishes a DT from a digital shadow (DS), which is only a pure assignment of the status data by a measurement to a specific asset at a specific instant in time (Bergs et al. 2021). This characteristic that DTs cover the entire product lifecycle is underlined in a current research project exploring the solutions for DTs for AI application. The underlying concept is visible in Fig. 6 (published with CC license in Arff et al. 2023).

The central element of this DT architecture is a semantically rich knowledge base. This knowledge base is represented as a knowledge graph, which may be generated applying the Design Compiler 43 (Neumaier et al. 2022), a central component of the Design Cockpit 43® by IILS mbH, Trochtelfingen, Germany. This system generates an output containing the ontology definition and product geometry. Present in the system are also execution and performance data which represent existing operation data from past operations. They include, for example, sensor signals, control system outputs and actuator commands. These data are also stored in the central knowledge graph which serves as a data basis for machine learning techniques and explainable AI (XAI) approaches (Arff et al. 2023). For enabling behaviour simulations, so-called functional mock-up units (FMUs – compare Larsen et al. 2016) are applied, which can also be

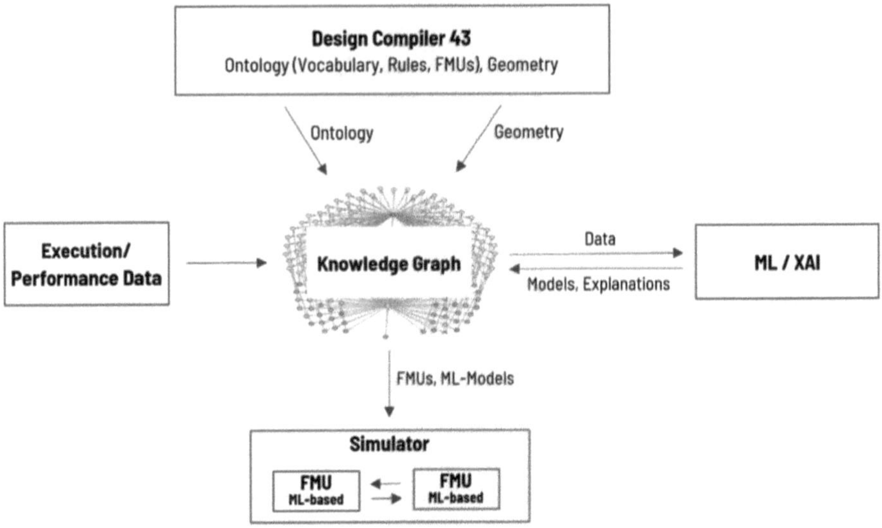

Fig. 6. Digital twin architecture for artificial intelligence application

AI-based. Based on this kind of DT architecture, which leads to a holistic AI application, it can be assumed that future design engineers need to fill a semantically rich knowledge base, i.e. to document geometry, system structures, decision rationale, etc. in a form this is generally appropriate for further processing by AI. Similar to new design paradigms, this may alter the design process considerably. Additionally, the direct application of AI in engineering design is currently starting – in simulation, but also design synthesis itself. Large language models (such as ChatGPT) are currently the focus of numerous discussions in society and industry. Design engineers are already using this kind of technology for communication and documentation purposes. Similar approaches are currently investigated by large producers of simulation software in order to replace complex simulations (such as finite element analysis (FEA) simulations) by searches for similar simulations which were done in the past and their results. Currently, it is hard to give a profound prognosis which future applications will be possible and sensible. Figure 7 shows a design engineer; this figure was created with AI using Bing Create.

Figure 7 can give an impression for the potential of AI in design; a closer look at the engine also indicates some limits and dangers.

As an overview result of the review concerning AI in design engineering, we can confidently say that the interest in data-based design methods and Explainable Artificial Intelligence (XAI) has increased in recent years. A current review reports that the use of AI methods in engineering design applications helps to obtain efficient, fast, accurate and comprehensive results (Yüksel et al. 2023). Which changes can engineering designers expect? Verganti et al. (2020) conclude that design practice, in the age of AI, is completely different, because the role of humans is not to develop full solutions (which are developed by AI), but to understand which innovation problems are meaningful, by framing the innovation effort, and by setting up the software, data infrastructure, and problem-solving loops that will solve them.

Fig. 7. Design engineer (figure created with AI using Bing Create)

2.5 Additional Domains, Requirements and Approaches

During the last decades again and again additional domains, requirements and approaches have expanded the problem and solution space for technical products. Additional domains frequently result from the general trend to include automatic control. New approaches result from innovations in the process tool chain if design engineers, e.g. systems for topology optimization. Additional requirements result from increasing competition as well as rising performance, quality, safety, reliability and sustainability expectations. Often the requirements are harder to meet, because of increasing system complexity. One example are faults which cannot be completely avoided in complex products leading to the necessity to develop fault-tolerant products. This can be supported by the methods and tools of fault-tolerant design (compare Stetter 2020b). Important elements of fault-tolerant design can be found in Fig. 8 (published under CC license in Stetter 2020c).

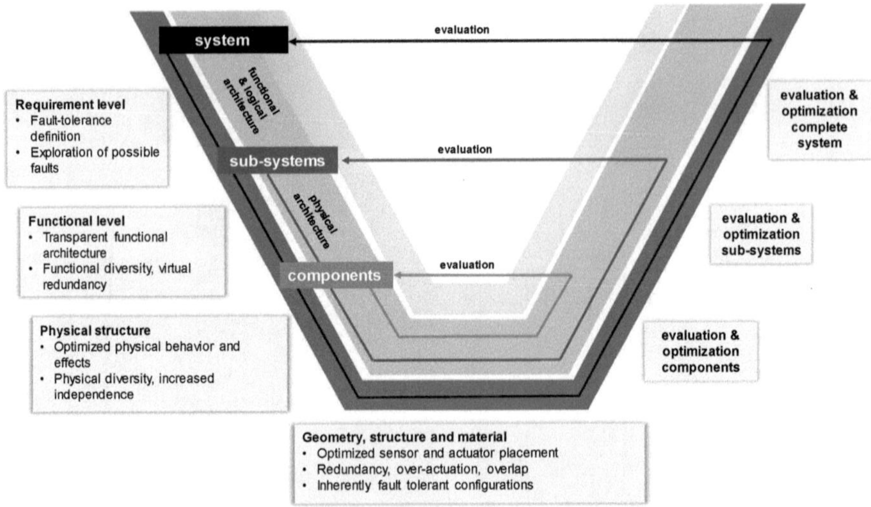

Fig. 8. Important elements of fault-tolerant design

The approaches to enhance the fault-tolerance of a technical system are ordered concerning a model of product concretization (compare Ponn & Lindemann 2011). On each level product improvement possibilities are listed together with tools and methods intended to support fault-tolerant design (Stetter 2020c). In spite of the incontestable appeal of supporting tools and methods for design engineers, the process may become more complex and new capabilities and knowledge may be required. Also, for effective and efficient use, all tools and methods have to be adapted and integrated in the overall process. Reviews at product development milestones have to assure that additional requirements are met. These reviews may become too long and a prioritization of the most important requirements may become an issue. The capability to keep an overview and to distinguish levels of importance may be getting more and more important for design engineers.

3 Drivers of Changes

This section analyses drivers for change in the daily work of design engineers, i.e. factors which may be underlying causes for a changed working environment and working style.

3.1 The Presence of a Digital Design Process Management

A combination of an implementation of MBSE and additional requirements and approaches may lead to an intensified digital design process management, e.g. by means of predefined process sections in a product data management (PDM) or product lifecycle management (PLM) system. The process management system may require design engineers to follow predefined working steps in synthesis (e.g. required product models with a respective level of maturity) and analysis (e.g. required simulations). The process management system may also contain mechanisms for version, variant and configuration management as well as product life cycle management. Initially, this kind of process management system may help design engineers to plan, execute and document their problem solving processes. However, a strict guidance through the course of the product development process will definitely change the daily work of design engineers and may even prevent creative solution finding.

3.2 Increased Possibilities for Fast Prototyping

In recent years, the daily work of design engineers was considerably altered by an increase in the possibilities for fast prototyping. Firstly, this relates to the virtual domain, because these days many design engineers are able to perform fast finite element method (FEM) analyses by themselves. This capability is caused by expanded education contents at Universities and by the integration of analysis modules within CAD programs. Secondly this relates to physical domain because of radically improved possibilities for 3-D-printing. Today, in most engineering departments, systems for additive manufacturing are present, which allow a quick and inexpensive realization of prototype parts. These two aspects are summarized in Fig. 9.

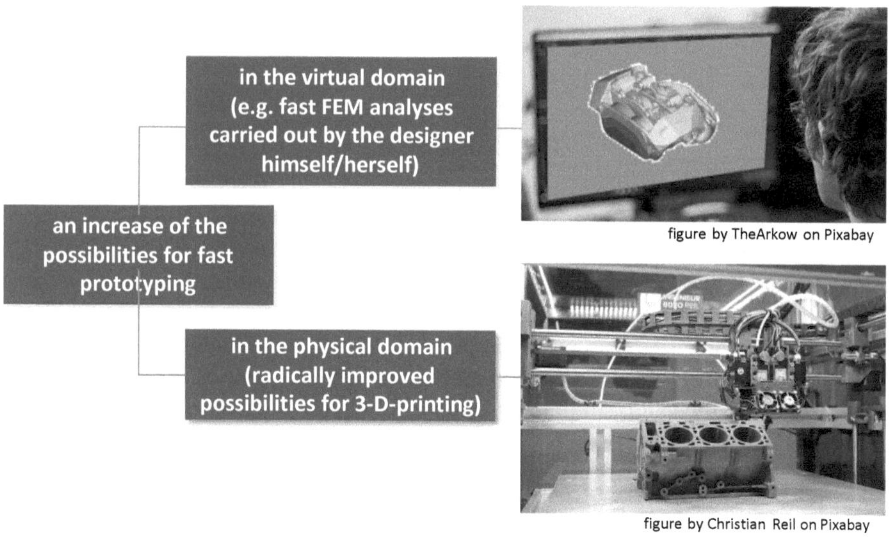

Fig. 9. Increased possibilities for fast prototyping

3.3 Altered Communication Behavior

Another important alteration of the daily work of design engineers was intensified in the last years by the COVID pandemic and the need for social distancing. More than ever before, design engineers use digital collaboration platforms and in some contexts virtual meetings have become the preferred mode of collaboration. It is an undeniable fact that this leads to certain improvements such as a reduction of inefficient travelling times and the possibility to include people in meetings whose attendance was not possible before. However, also some drawbacks became apparent in the last years e.g. concerning remote collegiality (Bleakley et al. 2022). A general consensus can be observed that team building is very difficult via remote communication as well as the development of a common understanding and the transfer of tacit knowledge. These drivers of changes are summarized in Fig. 10.

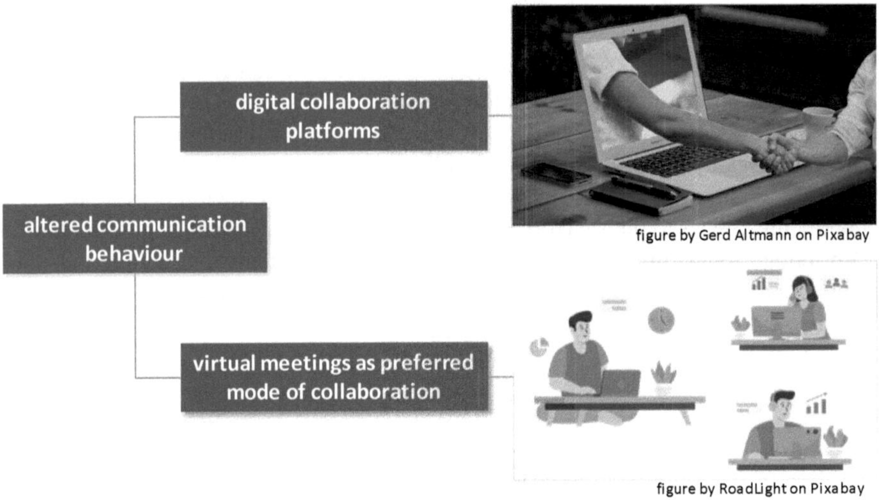

Fig. 10. Altered communication behaviour

Talks with design engineers in industrial companies have revealed that new communication media have had a significant impact on the way designers work, especially in the last three years. In many companies one to three home office days are possible for design engineers and virtual meetings are part of these days. New communication media or, more precise, an intensified and professional use of communication tools, which war already existing in most companies have enabled designer engineers to collaborate more easily and to communicate more effectively with more people. However, it is an undeniable fact that these communication tools have also created new challenges, such as ensuring data security as well as avoiding ongoing disturbances and information overload.

4 Consequences of Changes

The last two sections were concerned with changes of the working environment and working style of design engineers and alterations of their daily work. This section seeks to analyse some of the consequences, though this analysis is not yet exhaustive and needs to be complemented by future research.

One important consequence might be that complex, cyclic and agile processes are implemented, which are often more difficult to understand and to explain to new process members. This can be aggravated by the fact that digital processes contain less physical artefacts such as prototypes and test specimen and that the process is consequently more difficult to observe. Team members in a product development process might primarily collaborate remotely. It is probable that because of a lack of physical closeness certain opinions, assessments and uncertainties will not be transferred to other team members in an optimum manner. This situation might be especially difficult for new team members – they may have difficulties to observe activities of their co-workers, to learn from co-workers and to ask seemingly mundane questions.

Besides some negative consequences, a digital design process may have enormous advantages, because of the potential to capture a much larger share of the generated and applied information and to also capture the interconnections. Still, a central question may remain: do the changes lead to a design process which captures all essential ideas, issues, background information and process management elements digitally? Most probably the answer for now and the foreseeable future is no. This is caused by the immense amount of available formal and informal information. Additionally, this may be caused by the unique nature of design seeking for innovative products and solutions and may be caused by a refusal to share certain information.

In a nutshell, digital development processes combine certain advantages and drawbacks and are not yet developed and implemented in a manner that enables their full improvement potential. Still, return to the prior modes of working is very unlikely and most probably not sensible. Consequently, the next section will discuss implications for future design processes and design methods.

5 Implications for Future Design Processes and Design Methods

As stated above, a return to prior modes of working is unlikely. This section seeks to give an initial answer to the question how industrial companies and universities should design future design processes and design methods. Very probably, design processes in the next years and decades will be characterized by a mixture of digital and conventional processes, digital and conventional artefacts and digital and conventional communication media. One challenge will be to capture as much information as possible digitally in a form that allows machine interpretation. Current investigations show the necessity to work collaboratively, to identify the core aims and goals, as well as the necessary technologies, methodologies and innovations for creating a solid digital transformation strategy in didactics (Aleksandrov et al. 2020). Currently, powerful approaches are being developed such as machine learning and big data analysis, but they depend on the availability of appropriate data with multi-level structural information. One important characteristic of promising improvement approaches is that the concerned digital tools are easy to use and the "payback" for their application is imminent. One cornerstone of future product development processes will be a further expansion of product and process simulation capabilities. Data driven analysis approaches as a complement to physics driven analysis approaches may lead to interesting insights. Another promising field is the integration of process management approaches with neighbouring fields. For instance, MBSE might be combined with model-based safety analysis (MBSA) and reliability engineering approaches and digital quality assurance, e.g. digitally guided 8-D-processes.

Important aspects concerning future design processes are summarized in Fig. 11.

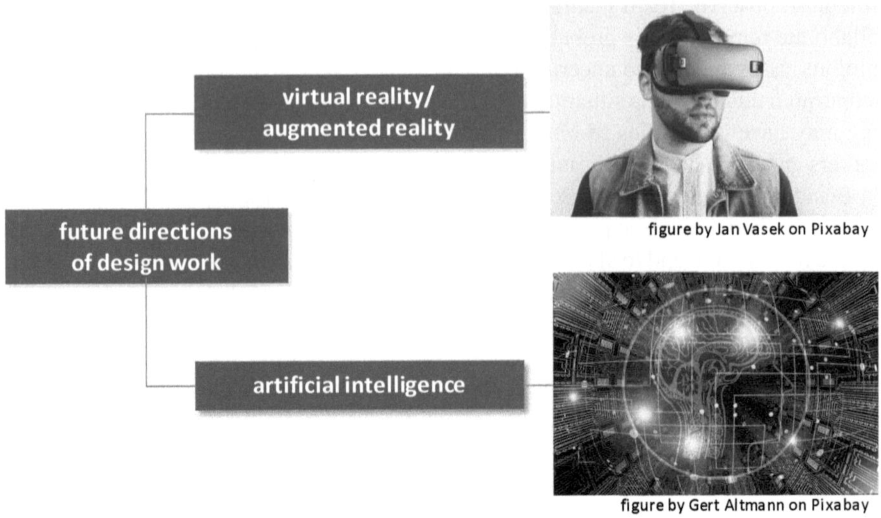

Fig. 11. Aspects of future design processes

It is probable that physical artefacts will still be important, but will be realized in later stages of the process with higher solution confidence.

Last but not least the changes in industry will lead to new education issues. Universities will need to answer the question how they can educate future design engineer according for the new working environment and working style without concentrating on some prominent trends with short lifetime. A promising attempt might be a combined effort by universities and industry. A future challenge for industrial companies can be the so-called "on-boarding" of new employees, because the experienced colleagues might be working at home and will not be present for quick questions and the observation of successful working styles. Future university graduates need to be able to work more independent from the beginning; bachelor and master theses in close cooperation with industry might be one promising approach for increasing such capabilities.

Initially, the question was asked, if future design processes will allow to capture all essential ideas, issues, background information and process management elements digitally. Is it then possible to automate this process completely? This article would conclude with the hypothesis that this is not possible – based on the unique nature of design seeking for innovative products and solutions.

Acknowledgement. Parts of the described research were funded by the Carl Zeiss Foundation in the scope of the project AI-based digital twin (KI-basierter digitaler Zwilling (KIDZ)).

References

Albert, A., Christian, Z.: Extending SysML for engineering designers by integration of the contact & channel–approach (C&C2-A) for function-based modeling of technical systems. Procedia Comput. Sci. **16**, 353–362 (2013)

Aleksandrov, A.A., Tsvetkov, Y.B., Zhileykin, M.M.: Engineering education: key features of the digital transformation. In: ITM Web of Conferences, vol. 35, p. 01001 (2020)

Arnold, P., Rudolph, S.: Bridging the gap between product design and product manufacturing by means of graph-based design languages. In: TMCE 2012 Symposium, Karlsruhe (2012)

Bleakley, A., et al.: Bridging social distance during social distancing: exploring social talk and remote collegiality in video conferencing. Hum. Comput. Interact. **37**(5), 404–432 (2022)

Benbya, H., Pachidi, S., Jarvenpaa, S.L.: Special issue editorial: artificial intelligence in organizations: implications for information systems research. J. Assoc. Inf. Syst. **22**(2), 281–303 (2021)

Berendsen, G., Beckett, R.C.: Quality 4.0 – innovation in digital ecosystem quality assurance. In: 21st International CINet (virtual) Conference "Practicing Continuous Innovation in Digital Ecosystem" Milan, Italy (2020)

Bergs, T., Gierlings, S., Auerbach, T., Klink, A., Schraknepper, D., Augspurger, T.: The concept of digital twin and digital shadow in manufacturing. Procedia CIRP **101**, 81–84 (2021)

Ehrlenspiel, K., Meerkamm, H.: Integrierte Produktentwicklung: Denkabläufe, Methodeneinsatz, Zusammenarbeit, 6, vollständig überarbeitete und erweiterte Auflage. München Wien (2017)

Feldmann, S., Vogel-Heuser, B.: Diagnose von Inkonsistenzen in heterogenen Engineering-Daten. Handbuch Industrie 4.0 Bd. 2: Automatisierung, pp. 315–334 (2017)

Gräßler, I., Wiechel, D., Pottebaum, J.: Role model of model-based systems engineering application. In: IOP Conference Series: Materials Science and Engineering, vol. 1097, p. 012003. IOP Publishing (2021)

Grauberger, P., Wessels, H., Gladysz, B., Bursac, N., Matthiesen, S., Albers, A.: The contact and channel approach–20 years of application experience in product engineering. J. Eng. Des. **31**(5), 241–265 (2020)

Hashmi, M.A., Mo, J.P.T., Becket, R.C.: Transdisciplinary systems approach to realization of digital transformation. Adv. Eng. Inform. **49**, 101316 (2021)

Holder, K., et al.: Model-based requirements management in gear systems design based on graph-based design languages. Appl. Sci. **7** (2017)

Jiao, R., Luo, L., Malmqvist, J., Summers, J.: New design: opportunities for engineering design in an era of digital transformation. J. Eng. Des. **33**(10), 685–690 (2022). https://doi.org/10.1080/09544828.2022.2147270

Lakemond, N., Holmberg, G., Pettersson, A.: Digital transformation in complex systems. IEEE Trans. Eng. Manag. (2021). https://doi.org/10.1109/TEM.2021.3118203

Larsen, P.G., et al.: Integrated tool chain for model-based design of cyber-physical systems: the INTO-CPS project. In: 2016 2nd International Workshop on Modelling, Analysis, and Control of Complex CPS (CPS Data), pp. 1–6. IEEE (2016)

Mandel, C., Schwarz, S., Brix, T., Albers, A., Husung, S.: Improving system of objectives maturity through systematic reuse of knowledge using ontology-based knowledge representations. In: Proceedings of the 60th Ilmenau Scientific Colloquium (2023)

Mhenni, F., Choley, J.Y., Penas, O., Plateaux, R., Hammadi, M.: A SysML-based methodology for mechatronic systems architectural design. Adv. Eng. Inform. **28**(3), 218–231 (2014)

Neumaier, M., Kranemann, S., Kazmeier, B., Rudolph, S.: Automated piping in an airbus A320 landing gear bay using graph-based design languages. Aerospace **9**, 140 (2022)

Ponn, J., Lindemann, U.: Konzeptentwicklung und Gestaltung technischer Produkte. Springer, Heidelberg (2011). https://doi.org/10.1007/978-3-642-20580-4

Sagodi, A., Engel, C., Schniertshauer, J., Van Giffen, B.: Becoming certain about the uncertain: how AI changes proof-of-concept activities in manufacturing–insights from a global automotive leader (2022)

Salehi, V., Florian, G., Taha, J., et al.: Implementation of systems modeling language (Sys-ML) in consideration of the consens approach. In: DS 92: Proceedings of the DESIGN 2018 15th International Design Conference, pp. 2987–2998 (2018)

Stetter, R.: Approaches for modelling the physical behavior of technical systems on the example of wind turbines. Energies **13**(8) (2020a)

Stetter, R.: Fault-Tolerant Design and Control of Automated Vehicles and Processes. Insights for the Synthesis of Intelligent Systems. Springer, Cham (2020b). https://doi.org/10.1007/978-3-030-12846-3

Stetter, R.: A fuzzy virtual actuator for automated guided vehicles. Sensors **20**(15) (2020c)

Trauer, J., Pfingstl, S., Finsterer, M., Zimmermann, M.: Improving production efficiency with a digital twin based on anomaly detection. Sustainability **13**(18), 10155 (2021)

Trauer, J., Schweigert-Recksiek, S., Engel, C., Spreitzer, K., Zimmermann, M.: What is a digital twin?—definitions and insights from an industrial case study in technical product development. In: Proceedings of the Design Society: DESIGN Conference, vol. 1, pp. 757–766 (2020)

Verganti, R., Vendraminelli, L., Iansiti, M.: Innovation and design in the age of artificial intelligence. J. Prod. Innov. Manag. **37**(3), 212–227 (2020)

Weilkiens, T.: Systems Engineering with SysML/UML: Modeling, Analysis, Design. Elsevier, Amsterdam (2011)

Winkler, M., et al.: Accelerating Automotive's AI Transformation: How Driving AI Enterprise-Wide Can Turbo-Charge Organizational Value. Capgemini Research Institute, 2019 (2019)

Yüksel, N., Börklü, H.R., Sezer, H.K., Canyurt, O.E.: Review of artificial intelligence applications in engineering design perspective. Eng. Appl. Artif. Intell. **118**, 105697 (2023). https://doi.org/10.1016/j.engappai.2022.105697

Distinguishing Between Product Development and Project Management in Agile and Conventional Practices

Lisa Siewert(✉) and Kilian Gericke

University of Rostock, Rostock, Germany
lisa.siewert@uni-rostock.de

Abstract. The distinction between planning a product development process and managing a development project highlights their interconnected roles in developing innovative solutions. Additionally, the often-confused terms methodology, method, and model are clarified, with their distinct roles in product development and project management delineated. Context factors that influence the suitability of agile or traditional working techniques are identified, providing insight into when each approach is most effective. These findings offer a basis for further exploration into the compatibility and integration of agile and traditional approaches in various project environments.

1 Introduction

Product development projects are characterized by significant levels of innovation, creativity, concurrency, and interaction. They are characterised by ambiguities, uncertainties, and interdependencies among tasks, outcomes, personnel, and tools, all of which contribute to the complexity of the development process (Kline 1985; Browning and Ramasesh 2007). Additionally, a multitude of factors influence project dynamics, such as company size, project management maturity, technological uncertainty, degree of innovation, project scope, and team size (Shenhar and Dvir 1996; Boehm 2004; Project Management Institute 2013; Turner and Ledwith 2018). Despite each project being unique, they share common elements, which have been the focus of researchers aiming to develop practices that support both project management and the product development process (Smith and Morrow 1999).

Wynn and Clarkson (2018) discuss various process models that have emerged from engineering design research, which cover different levels of abstraction. These models vary based on the characteristics of the specific development situation and the objectives they aim to fulfil. A wide range of concepts, methods, and tools are available to support the management of product development (Blessing 1995; Eisenbart, Gericke and Blessing 2011). As product complexity increases, the importance of these practices should rise correspondingly. However, it is still common to observe failures in development projects due to false application of methods (Portman 2021).

One challenge in applying the above named practices lies in the gap between academic research and industry application. Many companies remain unaware of these

K. Gericke et al. (Eds.): DSC 2022/2023, PTT, pp. 53–69, 2026.
https://doi.org/10.1007/978-3-032-16278-6_5

practices, often due to barriers in knowledge transfer (Araujo et al. 1996; Gericke et al. 2020). Additionally, misinterpretation or incorrect application of these practices can lead to project failures (Becker 2006; Albers et al. 2019). Conventional product development methodologies, often criticized for being too linear, rigid, and over-planned, have been identified as unsuitable for managing highly innovative or dynamic projects (Cooper 2014; Edwards et al. 2019). This has spurred the adoption of agile approaches, which offer greater flexibility and adaptability to rapidly changing requirements (Cho 2009; Gustavsson 2016; Schmidt et al. 2019).

The rise of agile methodologies has sparked comparisons with traditional approaches, focusing on their respective strengths and limitations, and exploring which approach might dominate future product development (Serrador and Pinto 2015; Albers et al. 2019; Heimicke et al. 2019). According to a study by Wysocki (2014), project managers often default to conventional practices, even though only 20% of development projects truly fit the conventional mold. Tiwana and Kail (2004) identified the use of an inappropriate methodology as a critical risk factor for project failure, emphasizing the importance of aligning the development approach with both the project's characteristics and its organizational context to enhance the likelihood of success (Tiwana and Keil 2004; Wysocki 2014). To address this challenge, some authors propose a hybrid approach, combining the strengths of both agile and conventional methodologies (Jovanovic and Beric 2018). This has led to the development of hybrid frameworks, which integrate elements of both paradigms to better navigate the complexities of modern product development (Cho 2009; Conforto and Amaral 2016; Cooper and Sommer 2016; Zasa, Patrucco and Pellizzoni 2020).

This paper seeks to further explore the comparison between agile and conventional practices and how they may be effectively combined by addressing the following research questions:

RQ1: What are the differences between planning a product development process and managing a development project?
RQ2: How can methodologies, methods, and models be differentiated?
RQ3: What characteristics determine whether a project is better suited for agile or conventional execution?

2 Study Design

To address research questions RQ1 and RQ2, the distinction between product development and project management was clarified by exploring the differences in their methodologies, models, and methods. While these domains often overlap, particularly at the level of practical applications, distinguishing them becomes clearer when precise definitions are identified for each concept. Definitions for "methodology," "method," and "model" were drawn separately from relevant literature, with Gericke et al. (2017) providing a foundational understanding of "design methodology" and "design methods" in product development. To extend this distinction into project management, a systematic review of literature using Scopus was conducted, ensuring a comprehensive exploration of these terms across both fields.

The search terms used included "project management product development," "project management method product development," "project management method literature review," and "project management methodology product development," yielding a total of 1,721 results. The analysis focused exclusively on publications related to product development, excluding fields such as construction projects and public event management. Additionally, studies that focused on the development of a single product or specialized areas like change management and risk management were also excluded. This filtering process resulted in a final set of 43 relevant publications, which were then analyzed in relation to the research questions. The references of these publications were also scanned for further related research.

This structured approach ensured that the literature review was comprehensive and focused on product development while maintaining relevance to the research questions.

3 The Relationship Between Product Development and Project Management Methodology, Process and Method

3.1 Differentiating Between Product Development and Project Management

Product Development

Product development refers to the structured process of transforming an initial technical problem, along with a set of objectives, into a fully realized solution (Ehrlenspiel and Meerkamm 2017; Gericke and Bender 2021). This process is commonly organized into four key phases: clarification of the task, conceptual design, embodiment design, and detail design.

The task clarification phase focuses on defining the project's objectives and identifying detailed requirements. During the conceptual design phase, a function structure is established, and various working principles and potential solution variants are developed. These variants are evaluated based on feasibility and effectiveness, with the preferred solution being selected. In the embodiment design phase, the product is broken down into modules, which are designed individually before being integrated into a cohesive whole. Finally, the detail design phase involves creating production documents, performing final testing, and ensuring readiness for manufacturing. This phased approach is common in engineering design processes, but it can vary slightly based on the specific field or type of product being developed (Gericke and Bender 2021).

Project Management

Project management is the application of knowledge, skills, tools, and techniques to achieve the specific goals of a project (PMI 2017b). It is generally divided into four phases: project initiation, project planning, project implementation and controlling, and project completion.

The project initiation phase involves developing a project charter and identifying stakeholders, laying the foundation for the project's objectives and scope.

During the planning phase, work packages are defined, a project structure plan is developed, and schedules, resource plans, and risk management strategies are created. This phase also includes budgeting. The implementation and controlling phase involves executing the project plan while monitoring progress. This includes leadership, governance, visual planning, and earned value management to ensure project alignment with goals. The project completion phase focuses on formally closing the project by conducting post-project reviews, capturing lessons learned, and finalizing documentation (Oehmen 2021).

The product development process and the management of a project both cover different sets of activities and generate different results. They take place in parallel and have to be considered separately. Figure 1 visualizes the correlation between the product development process and the project management phases in the context of a product lifecycle.

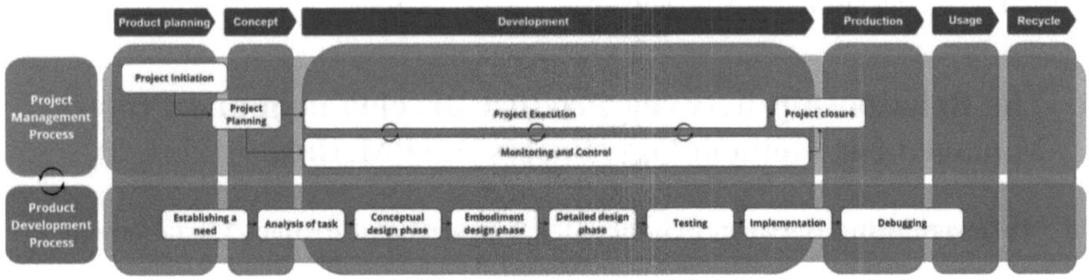

Fig. 1. Correlation between Product Development and Project Management

Key Differences between Product Development and Project Management

While both product development and project management contribute to bringing a product to market, they encompass different sets of activities and generate distinct results. Product development is focused on solving technical challenges and progressing through stages of design and testing, while project management is concerned with overseeing the entire project lifecycle, ensuring that different goals are met.

Both processes often run in parallel but are separate in scope and purpose. Product development concentrates on technical and functional requirements, whereas project management focuses on organizational and logistical coordination (Gericke and Bender 2021). It is essential to understand this distinction, as successful projects depend on the effective integration of both.

Recent Research on Integration of Product Development and Project Management

Recent research highlights the growing need for integration between product development processes and project management methodologies. Studies indicate that many project failures can be attributed to poor alignment between technical development and management practices Additionally, the increased complexity of modern products, often involving multidisciplinary teams and cutting-edge technologies, the need to synchronize product development and project management arose (Kapsali 2011; Oehmen et al. 2014). Hybrid approaches that combine elements of both agile and traditional methodologies have gained attention as a solution, particularly in industries where innovation and flexibility are critical. These hybrid methods emphasize iterative development, continuous stakeholder involvement, and adaptive planning, providing a more dynamic way to manage complex product development projects (Conforto and Amaral 2016; Cooper and Sommer 2016). These methodologies promote early-stage collaboration and help reduce the risk of miscommunication between engineering teams and project managers.

4 Differentiating Between Design Methodology and Project Management Methodology

Design Methodology

A design methodology refers to a systematic, procedural approach that outlines a structured plan of action for designing technical systems, based on established good practices. These methodologies typically organize activities into phases, specifying the sequencing and relationships between tasks. They guide designers by providing recommended methods, information artefacts, and decision-making priorities (Eisenbart, Gericke and Blessing 2011; Gericke and Blessing 2011; Gericke, Eckert and Stacey 2022). Design methodologies aim to rationalize the inherently creative nature of design work, facilitate planning, and improve communication across multidisciplinary teams (Gericke and Blessing 2011). Examples of design methodologies include (Gericke and Bender 2021):

- VDI 2221 (VDI2221 2019)
- VDI 2206, also known as the V-Model (VDI/VDE 2206 2021)
- Integrated Product Development by Andreasen and Hein (2000)
- Product Design and Development by Ulrich, Eppinger, and Yang (2020)

Project Management Methodology

A project management methodology is a comprehensive framework of best practices, tools, and techniques aimed at achieving project goals and monitoring progress against the plan (Chin, Hwa Yap and Spowage 2010; International Project Management Association 2016; Gericke and Bender 2021). These methodologies serve as roadmaps for navigating the project lifecycle, encompassing initiation, planning, execution, and monitoring. They provide standard terminology, risk management frameworks, and guidelines for project governance (Chin, Hwa Yap and Spowage 2010; Chin, Spowage and Yap 2012).

The overarching goal of a project management methodology is to increase the likelihood of project success, ensure quality, and maintain control through defining tasks, assigning resources, and continuously adjusting plans (Harold 2009; International Project Management Association, 2016; Radujković and Sjekavica 2017). This is realised through defining the required work, establishing the extent of work, allocating the required resources as well as constant planning, monitoring and adjustment (Munns and Bjeirmi 1996). A project management methodology provides a team with a set of processes that can be scaled or substituted and assists a team in understanding their scope of work, what they need to accomplish, how their work fits in and contributes to the project (Chin, Hwa Yap and Spowage 2010). Project management methodologies cover knowledge areas and process groups for which different methods are proposed (Timinger, Schmidtner and Reiche 2022). Chin and Spowage (Chin, Hwa Yap and Spowage 2010) classify project management methodologies in five levels:

- level one: best practices, standards, guides (e.g. PMI, PRINCE2)
- level two: sector-specific methodologies (e.g. Scrum)
- level three: organizational methodologies (e.g. Microsoft Solution Framework, Rational Unified Process, PROPS)

- level four: project-specific methodologies
- level five: individual methodologies

4.1 Differentiating Between Design Process Model and Project Management Model

A design process model provides a framework to ensure proper execution of technical product development. It encompasses activities like requirements management, implementation, and testing (Timinger, Schmidtner and Reiche 2022). These models may be activity-based, stage-based, problem-oriented, or solution-oriented, and they require interpretation and adaptation (Wynn and Clarkson 2005). A typical design process model involves phases such as (Gericke and Blessing 2011; Sommer, Dukovska-Popovska and Steger-Jensen 2014):

- Establishing a need
- Task analysis
- Conceptual design
- Embodiment design
- Detailed design
- Implementation

Procedural models aim to visualize the development process to support management, improve understanding, and enhance communication across teams (Browning and Ramasesh 2007; Sommer, Dukovska-Popovska and Steger-Jensen 2014). Examples include the Stage-Gate-Model by Cooper (Cooper 1990) and the spiral development process by Boehm (Boehm 2000).

A project management model, on the other hand, is a simplified representation of reality designed to optimize processes or provide strategic guidance. These models reflect strategies and frameworks, helping teams navigate project complexities. Examples include the OSCAR model, the Theory of Needs, and the Stacey Matrix (PMI 2017b).

4.2 Differentiating Between Design Method and Project Management Method

Design Method

Design methods are structured procedures that guide how tasks should be performed and how outcomes are achieved. These methods require interpretation based on the environment in which they are applied (Andreasen 2011). A design method specifies what information is required, how tasks are decomposed, and how they should be sequenced (Gericke, Eckert and Stacey 2017). Examples include use value analysis and affinity mapping (Gericke, Roschuni and Kramer 2015). The purpose of design methods is to reach specific goals, solve problems, or express knowledge (Roozenburg and Cross 1991).

Project Management Method

Project management methods are techniques or procedures used to achieve project objectives or create deliverables. These methods help estimate data, support project execution, and structure tasks (PMI 2017b; Albertos and Ribeiro 2022). Examples include the Critical Path Method, SWOT analysis, Daily Standups, and Lessons Learned (Pons 2008; PMI 2017b).

Gantt charts, Kanban boards, and work-breakdown structures are commonly used artefacts generated through these methods (Eckert and Clarkson 2010).

In conclusion, design methodologies and project management methodologies serve distinct but complementary roles in product development. Design methodologies focus on solving technical challenges and guiding the creative process, while project management methodologies focus on controlling and monitoring the overall project. Proper integration of both, combined with clear method use, enhances the chances of project success. Recent literature suggests that hybrid approaches, which combine the best of both traditional and agile methodologies, can further improve outcomes in complex projects (Conforto and Amaral 2016; Cooper and Sommer 2016a).

5 Differentiating Between Conventional and Agile Product Development

5.1 Characteristics of Conventional and Agile Product Development

Conventional (Plan-Driven) Development

Plan-driven, or conventional product development, is typically characterized by a structured, linear approach where tasks are predefined and executed in sequential phases. This method emphasizes early decision-making to avoid costly changes later in the development cycle. As a result, plan-driven approaches tend to limit the range of possibilities as the project progresses, aiming to lock down requirements early and minimize modifications later in the process (Wynn and Clarkson 2018).

Key features of conventional development include:

- Simplicity and ease of management: Phases are clearly defined, and there is minimal iteration between them.
- Predictability: Tasks and timelines are planned in detail, enabling clear expectations and milestones.
- Risk of late-stage issues: Since users typically see results only at the end, errors or mismatches with requirements might be discovered too late in the process, resulting in significant rework or delays (Heimicke, Chen and Albers 2020).
- Limited flexibility: The rigid structure limits the ability to adapt to changes once development has started, potentially resulting in a final product that is not fully aligned with user needs or market conditions

Agile Product Development

Agile development is flexible, lightweight, and iterative. It emphasizes continuous user feedback and adaptation to changing requirements throughout the development cycle. Agile methods embrace change as an integral part of adding value to the project. Key decisions and requirements are revisited iteratively, allowing for more responsiveness to stakeholder input and evolving project circumstances (Boehm 2004; Conboy 2009; Serrador and Pinto 2015).

Key features of agile development include:

- Iterative and incremental: Work is divided into smaller cycles (sprints) where features are developed, tested, and improved in rapid succession.
- User involvement: Stakeholders are regularly engaged in prioritizing and validating features, ensuring that the development process aligns with their evolving needs.
- Embracing change: Agile methodologies are designed to accommodate changes throughout the project lifecycle, whether due to shifting priorities, user feedback, or new discoveries.
- Self-organizing teams: Agile relies on decentralized team structures where members are empowered to make decisions and adjust processes dynamically during the project.

Comparatively, agile methodologies are better suited for complex projects where requirements are uncertain or may change during development. Conventional methodologies work well in projects with well-defined goals and minimal expected variability.

5.2 Context Factors that Influence the Scalability of Project Management Approaches

The choice between conventional and agile development is influenced by several contextual factors. Research into the success of agile and conventional methods across different industries has identified key environmental and project-specific variables that affect the effectiveness of each approach. (Mule and Hehenberger 2020; Bianchi et al. 2022; Rehder et al. 2023). Table 1 highlights these factors as recommended reported in current literature.

Table 1. Context factors of conventional and agile practices

		Conventional	Agile
Application	Primary Goals	• Predictability, stability, high assurance (Boehm 2004)	• Rapid value, responding to change (Boehm 2004)
	Fundamental assumption/ Environment	• Systems are fully specifiable, predictable and are built through meticulous and extensive planning (Dybå and Dingsøyr 2008) • Stable environment characterized by high predictability (Cooper and Sommer 2018; Zasa, Patrucco and Pellizzoni 2020) • Stable, low change, project/organization focused (Boehm 2004)	• High quality adaptive software is developed by small teams using the principles of continuous design improvement and testing based on rapid feedback and change (Dybå and Dingsøyr 2008) • Turbulent environment characterized by high uncertainty (Cooper and Sommer 2018; Zasa, Patrucco and Pellizzoni 2020) • Turbulent, high change, project focused (Boehm 2004)

(*continued*)

Table 1. (*continued*)

		Conventional	Agile
	Size	• Larger teams and projects (Boehm 2004)	• Smaller teams and projects (Boehm 2004)
	Development approach	• Life-cycle model (Dybå and Dingsøyr 2008) • Plan-driven process based on extensive control of sequential phases (Cooper and Sommer 2018; Zasa, Patrucco and Pellizzoni 2020)	• The evolutionary delivery model (Dybå and Dingsøyr 2008) • Iterative development process based on feedback and change (Cooper and Sommer 2018; Zasa, Patrucco and Pellizzoni 2020)
	Success	• Focus on time and budget	• Focus on satisfied customer
Management	Style/Decision making	• Command and control (Dybå and Dingsøyr 2008) • Key strategic and operational decisions taken at predefined gates (Cooper and Sommer 2018; Zasa, Patrucco and Pellizzoni 2020) • Managerial decision making (Vinekar, Slinkman and Nerur 2006)	• Leadership and collaboration (Dybå and Dingsøyr 2008) • Key strategic and operational decisions taken by team members (Cooper and Sommer 2018; Zasa, Patrucco and Pellizzoni 2020) • Mluralistic decision making (Vinekar, Slinkman and Nerur 2006)
	Knowledge management	• Explicit (Boehm 2004; Dybå and Dingsøyr 2008)	• Tacit (Boehm 2004; Dybå and Dingsøyr 2008)
	Desired organizational structure	• Mechanistic (bureaucratic with high formalization) • Aimed at large organizations (Dybå and Dingsøyr 2008)	• Organic (flexible and participative encouraging cooperative social action) • Aimed at small and medium sized organizations (Dybå and Dingsøyr 2008)
	Planning and control	• Traditional planning tools such as Gantt charts, milestones and critical path planning (Cooper and Sommer 2018; Zasa, Patrucco and Pellizzoni 2020) • Documented plans, quantitative control (Boehm 2004)	• Real-time planning of sprints lasting 2–4 weeks (Cooper and Sommer 2018; Zasa, Patrucco and Pellizzoni 2020) • Internalized plans, qualitative control (Boehm 2004)
Interactions	Communication	• Formal (Dybå and Dingsøyr 2008) • Cooperation (Schmidt et al. 2019)	• Informal (Dybå and Dingsøyr 2008) • Collaboration (Schmidt et al. 2019)
	Customer relations	• As-needed customer interactions • Focused on contract provisions • Customers not always collocated (Boehm 2004)	• Dedicated on-site customers • Focused on prioritized increments • Collocated customers (Boehm 2004)

(*continued*)

Table 1. *(continued)*

		Conventional	Agile
	Culture	• Comfort and empowerment via framework of policies and procedures (Boehm 2004) • Individual reward systems (Vinekar, Slinkman and Nerur 2006)	• Comfort and empowerment via many degrees of freedom (Boehm 2004) • Team reward systems (Vinekar, Slinkman and Nerur 2006)
Technical	Uncertainty	• Attempt to control up-front to avoid later changes • More related to situations involving risks not uncertainty (de Vasconcelos Gomes et al. 2021)	• More appropriated to uncertain situations in which it is required to accommodate changes over time (de Vasconcelos Gomes et al. 2021)
	Requirements	• Formalized project, capability, interface, quality, foreseeable evolution requirements (Boehm 2004)	• Prioritized informal stories and test cases; undergoing unforeseeable change (Boehm 2004)
	Development	• Extensive design; longer increments; refactoring assumed expensive (Boehm 2004) • Life-cycle development (Vinekar, Slinkman and Nerur 2006)	• Simple design; short increments; refactoring assumed inexpensive (Boehm 2004) • Evolutionary development (Vinekar, Slinkman and Nerur 2006)
	Quality control	• Documented test plans and procedures (Boehm 2004) • Heavy planning and strict control • Heavy testing (Dybå and Dingsøyr 2008)	• Executable test cases define requirements (Boehm 2004) • Continuous control of requirements, design and solutions • continuous testing (Dybå and Dingsøyr 2008)

Conventional development is most effective when requirements are clearly defined from the outset, the environment is stable, and there is low tolerance for risk or change. This approach is common in industries where regulatory compliance or high levels of precision are critical, such as aerospace or civil engineering.

Agile development excels in environments characterized by high complexity and uncertainty, where rapid iterations and continuous feedback are essential to meet user needs. It is often applied in industries like software development, where requirements frequently evolve and customer feedback plays a central role in shaping the final product.

6 Summary and Conclusion

This paper differentiates between designing a product development process and managing a development project, highlighting their interdependent functions in innovating and executing solutions. Additionally, the study explores the frequently conflated terms—methodology, method, and model—to provide clarity and define their respective roles within product development and project management contexts. Drawing from these insights, the paper presents a set of contextual factors determining a project's suitability for agile or traditional methodologies. Furthermore, it establishes a groundwork for future research exploring the synergy between agile and traditional practices and their potential integration.

6.1 Difference Between Product Development and Project Management

The activities of product development and project management are interconnected, yet there is a distinct difference. Product development focuses on the technical aspects of creating a product, moving from initial problem-solving to a finalized solution. (Ehrlenspiel and Meerkamm 2017; Gericke and Bender 2021). It is concerned with defining the technical requirements, developing solution variants, and detailing the design for production. Each of these stages is sequential and feeds into the next, with the ultimate goal of creating a physical or digital product. On the other hand, project management focuses on organizing and controlling the activities necessary to complete a project within scope, time, and budget constraints. Project management encompasses activities such as project initiation, planning, implementation and control, and project closure. It provides the framework for ensuring that resources are allocated properly, timelines are met, and risks are managed (PMI 2017a; Oehmen 2021).

Aside from the activities product development and project management entail, they also differ in activities, responsibilities and required skill sets. Product development is about the product that is created. It is driven by creativity and problem-solving and requires technical expertise. Project management on the other hand is about how the product creation is organized and executed. To ensure that the product development process is completed successfully, planning, coordination and oversight are required. Project managers need to coordinate activities and stakeholders, balance time, goals and budget and need people skills to manage and motivate team members.

Although product development and project management are interdependent and often run in parallel, they aim for different outcomes and therefore require distinct management approaches. However, in practice, these two disciplines are frequently conflated. It is common for the same individual to model the product development process while managing the development project, which can blur the boundaries between the two. This overlap can lead to inappropriate use of methodologies—such as applying product development methods to project management—which results in criticism of process models being too broad to effectively guide project planning and decision-making (Gericke and Blessing 2011). Conversely, project management methodologies like Scrum are sometimes stretched to encompass product development tasks, despite being critiqued for lacking the detail needed to address specific development activities. This misapplication of tools designed for one domain to another underscores the need for clearer distinctions and tailored approaches for both product development and project management.

6.2 Difference Between Methodology, Method and Model

The terms methodology, method, and model are often used interchangeably, but they have distinct meanings in the context of both design and project management.

Methodology refers to a comprehensive and structured approach that outlines the steps and procedures necessary to complete a task or solve a problem. In product development, a design methodology organizes the phases and activities required to develop a product, such as requirement gathering, conceptual design, and testing (Gericke, Eckert and Stacey 2017). Similarly, in project management, a management methodology provides a framework for planning, executing, and monitoring a project, and it often includes best practices, tools, and techniques (Chin, Hwa Yap and Spowage 2010). Methodologies aim to standardize processes and ensure that tasks are completed efficiently.

A model represents a simplified or structured view of a complex process, often used to conceptualize or explain how a process works. In design, process models such as the Stage-Gate Model (Cooper 1990) or spiral development (Boehm 2000) represent a sequence of activities, helping to visualize and understand the flow of product development (Browning and Ramasesh 2007). In project management, models like the Stacey Matrix provide a framework for understanding project complexity and determining the appropriate approach (PMI 2017a).

A method is a more specific procedure or technique used to perform a particular task. In product development, design methods help engineers solve design problems by prescribing how tasks are executed (Gericke, Eckert and Stacey 2022). Methods are practical tools that guide individual actions, such as use value analysis or affinity mapping (Gericke, Roschuni and Kramer 2015). In project management, methods like the Critical Path Method or SWOT analysis provide specific ways to manage tasks and assess project status (PMI 2017a).

A methodology, a model, a method and an artefact (any kind of outcome when applying a method, e.g. a gantt-chart) can be seen as a hierarchy (Fig. 2) where the specialization and the focus on a specific aspect of the product development increases as the level decreases. This can be described as a shift from a normative to an operative perspective. Product development and project management need to be balanced in order to create the right product and to successfully manage the development project.

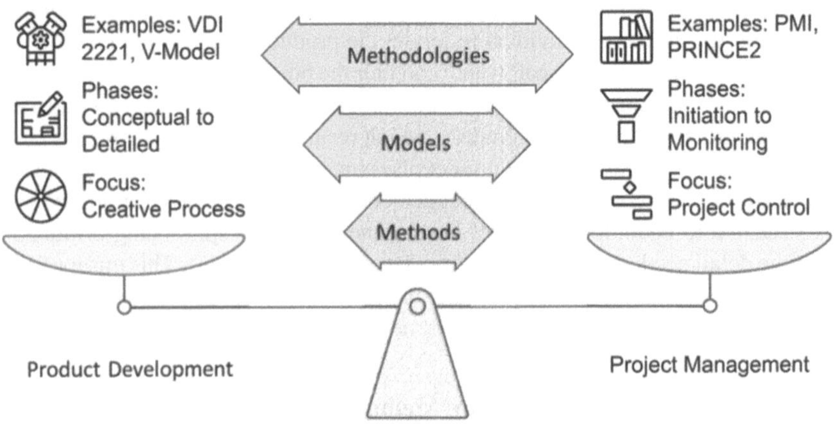

Fig. 2. Balancing product development and project management for successful outcomes

6.3 Suitability of Agile and Conventional Project Execution

The suitability of a project for either agile or traditional execution is influenced by several factors, as discussed in the literature. These factors include the complexity of the project, the stability of requirements, the level of stakeholder involvement, and the organization's risk tolerance (Mule et al. 2021; Rehder et al. 2023).

Project Complexity: Agile methodologies are particularly suited for high-complexity projects where uncertainty is high and requirements may evolve over time. In such contexts, the iterative nature of agile allows for continuous feedback and adaptation (Boehm 2004). On the other hand, traditional (plan-driven) methodologies are better suited for low-complexity projects where the outcomes are predictable and the scope is well-defined from the beginning (Wynn and Clarkson 2018).

Requirement Stability: Projects with stable and well-defined requirements are best suited for traditional methodologies, as the plan-driven approach assumes that changes are minimized once the project is in motion. In contrast, agile methods excel in environments where requirements are likely to change, as they accommodate flexibility and iteration throughout the development process (Serrador and Pinto 2015).

Stakeholder Involvement: Agile methods thrive when stakeholders can provide continuous feedback and prioritize changes. This high level of engagement ensures that the product aligns with user needs. In traditional approaches, stakeholder involvement is typically limited to the initial planning phases, with minimal opportunity for feedback during execution (Conboy 2009).

Risk Tolerance: Agile methodologies are more appropriate in situations where there is a higher tolerance for risk and uncertainty. The agile framework assumes that changes will occur and views them as opportunities for improvement. Traditional methodologies, on the other hand, aim to minimize risk by locking down requirements early and managing change cautiously (Heimicke, Chen and Albers 2020).

In conclusion, the choice between agile and traditional methodologies depends on these contextual factors. Agile methods are generally better for projects with high uncertainty, dynamic requirements, and the need for continuous user engagement, while traditional methods are more suitable for predictable, low-complexity projects with stable requirements.

The findings of this paper show that the Stage-Gate-Model and Scrum cannot be compared to each other in contrary to many reviews that can be found in the literature. This chapter argues that a comparison of product development approaches and project management approaches require care and a clearly articulated focus of the comparison, i.e. at what level these approaches are compared (methodology, process, model or method). Conventional process models like the Stage-Gate-Model are criticised for being too general to help project planning and daily decisions (Browning 2009) and their product focused perspective are found to be excluding management aspects (Blessing 1995). These objections are not wrong, but it is not the intention of a process model to serve as a project management practice, even though they are often used as such in industry (Browning and Ramasesh 2007; Sommer, Dukovska-Popovska and Steger-Jensen 2014). On the other hand, people are sceptical with using agile management methodologies because of their lacking process visibility (Gustavsson 2016). Again, this can be traced back to the character of a project management methodology, which does not include a visualization of the development process. Therefore, both is needed to enable a successful product development and a successful project management. To further improve the understanding and the implementation of both practices, it has to be discussed, how practices for developing a product and practices for managing projects interact and how they can be combined.

6.4 Conclusion

In conclusion, establishing clear and universally accepted definitions for methodology, model, and method, as well as distinguishing between product development and project management, is essential. Standardized definitions ensure that both organizations and researchers operate with a common understanding, which enhances communication and clarity across industries. This uniformity allows for effective comparison of methodologies, models, and methods, helping to identify best practices and improve processes. Furthermore, consistent definitions support the creation of training programs and certifications, equipping professionals with the skills needed to drive better project outcomes. For research, these standards enable more rigorous, comparable studies, fostering collaboration and advancement in the field. Ultimately, clear definitions empower organizations to confidently implement methodologies and models, enhancing predictability and success in both product development and project management.

References

Albers, A., Heimicke, J., Müller, J., Spadinger, M.: Agility and its features in mechatronic system development: a systematic literature review. In: ISPIM Conference Proceedings (2019). (Preprint)

Andreasen, M.M.: 45 years with design methodology. J. Eng. Des. **22**(5) (2011). https://doi.org/10.1080/09544828.2010.538040

Andreasen, M.M., Hein, L.: Integrated product development. Institute for Product Development, Technical University of Denmark, Lyngby (2000). (Preprint)

Araujo, C.S., Benedetto-Neto, H., Campello, A.C., Segre, F.M., Wright, I.C.: The utilization of product development methods: a survey of UK industry. J. Eng. Des. (1996). https://doi.org/10.1080/09544829608907940

Becker, B.: Re-thinking the stage-gate process - a reply to the critics. Manag. Roundtable (2006). (Preprint)

Bianchi, M.J., Conforto, E.C., Rebentisch, E., Amaral, D.C., Rezende, S.O., De Padua, R.: Recommendation of project management practices: a contribution to hybrid models. IEEE Trans. Eng. Manag. **69**(6) (2022). https://doi.org/10.1109/TEM.2021.3101179

Blessing, L.: Comparison of design models proposed in prescriptive literature. The role of design in the shaping of technology (1995)

Boehm, B.: Spiral Development: Experience, Principles, and Refinements. Spiral Development Workshop, 9 February 2000, July 2000. (Preprint)

Boehm, B.: Balancing agility and discipline: a guide for the perplexed. In: Ramamoorthy, C.V., Lee, R., Lee, K.W. (eds.) SERA 2003. LNCS, vol. 3026, p. 1. Springer, Heidelberg (2004). https://doi.org/10.1007/978-3-540-24675-6_1

Browning, T.R.: The many views of a process: toward a process architecture framework for product development processes. Syst. Eng. **12**(1), 69–90 (2009). https://doi.org/10.1002/sys.20109

Browning, T.R., Ramasesh, R.V.: A survey of activity network-based process models for managing product development projects. Prod. Oper. Manag. **16**(2) (2007). https://doi.org/10.1111/j.1937-5956.2007.tb00177.x
Chin, C., Hwa Yap, E., Spowage, A.C.: Defining & Classifying Project Management Methodologies, XII, June 2014, pp. 1–9 (2010). https://www.researchgate.net/publication/233835849
Chin, C., Spowage, A., Yap, E.: Project management methodologies: a comparative analysis. J. Adv. Perform. Inf. Value **4**(1) (2012). https://doi.org/10.37265/japiv.v4i1.102
Cho, J.: A hybrid software development method for large-scale projects: rational unified process with scrum. Issues Inf. Syst., 340–348 (2009). https://doi.org/10.48009/2_iis_2009_340-348
Conboy, K.: Agility from first principles: reconstructing the concept of agility in information systems development. Inf. Syst. Res. **20**(3) (2009). https://doi.org/10.1287/isre.1090.0236
Conforto, E.C., Amaral, D.C.: Agile project management and stage-gate model—a hybrid framework for technology-based companies. J. Eng. Technol. Manag. JET-M **40** (2016). https://doi.org/10.1016/j.jengtecman.2016.02.003
Cooper, R.G.: Stage-gate systems: a new tool for managing new products. Bus. Horiz. **33**(3), 44–54 (1990). https://doi.org/10.1016/0007-6813(90)90040-I
Cooper, R.G.: What's next? After stage-gate. Res. Technol. Manag. **57**(1) (2014). https://doi.org/10.5437/08956308X5606963
Cooper, R.G., Sommer, A.F.: The agile–stage-gate hybrid model: a promising new approach and a new research opportunity. J. Prod. Innov. Manag. **33**(5) (2016). https://doi.org/10.1111/jpim.12314
Cooper, R.G., Sommer, A.F.: Agile–stage-gate for manufacturers. Res. Technol. Manag. **61**(2) (2018). https://doi.org/10.1080/08956308.2018.1421380
Dybå, T., Dingsøyr, T.: Empirical studies of agile software development: a systematic review. Inf. Softw. Technol. (2008). https://doi.org/10.1016/j.infsof.2008.01.006
Edwards, K., Cooper, R.G., Vedsmand, T., Nardelli, G.: Evaluating the agile-stage-gate hybrid model: experiences from three SME manufacturing firms. Int. J. Innov. Technol. Manag. **16**(8) (2019). https://doi.org/10.1142/S0219877019500482
Ehrlenspiel, K., Meerkamm, H.: Methodik der Integrierten Produkterstellung im Unternehmen. In: Integrierte Produktentwicklung (2017). https://doi.org/10.3139/9783446449084.004
Eisenbart, B., Gericke, K., Blessing, L.: A framework for comparing design modelling approaches across disciplines. In: ICED 11 - 18th International Conference on Engineering Design - Impacting Society Through Engineering Design (2011)
Gericke, K., Bender, B.: Pahl/Beitz Konstruktionslehre, Pahl/Beitz Konstruktionslehre (2021). https://doi.org/10.1007/978-3-662-57303-7
Gericke, K., Blessing, L.: Comparisons of design methodologies and process models across disciplines: a literature review. In: ICED 11 - 18th International Conference on Engineering Design - Impacting Society Through Engineering Design (2011)
Gericke, K., et al.: Supporting designers: moving from method menagerie to method ecosystem. Des. Sci. **6** (2020). https://doi.org/10.1017/dsj.2020.21
Gericke, K., Eckert, C., Stacey, M.: What do we need to say about a design method? In: Proceedings of the International Conference on Engineering Design, ICED (2017)
Gericke, K., Eckert, C., Stacey, M.: Elements of a design method - a basis for describing and evaluating design methods. Des. Sci. **8** (2022). https://doi.org/10.1017/dsj.2022.23
Gericke, K., Roschuni, C., Kramer, J.: Discovery and evaluation of design methods in practice: an empirical study. In: Proceedings of the ASME Design Engineering Technical Conference (2015). https://doi.org/10.1115/DETC201547387
Gustavsson, T.: Benefits of agile project management in a non-software development context - a literature review. In: Project Management Development – Practice and Perspectives: Fifth International Scientific Conference on Project Management in the Baltic Countries, April 14–15, Riga. University of Latvia, April 2016. (Preprint)

Harold, K.: Project Management – A Systems Approach to Planning, Scheduling, and Controlling, Patent Project Management (2009)

Heimicke, J., Chen, R., Albers, A.: Agile meets plan-driven - hybrid approaches in product development: a systematic literature review. In: Proceedings of the Design Society: DESIGN Conference (2020). https://doi.org/10.1017/dsd.2020.259

Heimicke, J., Niever, M., Zimmermann, V., Klippert, M., Marthaler, F., Albers, A.: Comparison of existing agile approaches in the context of mechatronic system development: potentials and limits in implementation. In: Proceedings of the International Conference on Engineering Design, ICED (2019). https://doi.org/10.1017/dsi.2019.226

International Project Management Association: Project Excellence Baseline for Achieving Excellence in Projects and Programmes, Ipma (2016)

Jovanovic, P., Beric, I.: Analysis of the available project management methodologies. Manag. J. Sustain. Bus. Manag. Solut. Emerg. Econ. **23**(3) (2018). https://doi.org/10.7595/management.fon.2018.0027

Kapsali, M.: Systems thinking in innovation project management: a match that works. Int. J. Proj. Manag. **29**(4) (2011). https://doi.org/10.1016/j.ijproman.2011.01.003

Kline, S.J.: Innovation is not a linear process. Res. Manag. **28**(4) (1985). https://doi.org/10.1080/00345334.1985.11756910

Mule, S., Hehenberger, P.: An empirical study on the usage of agile methods in the mechatronics industry. In: 2020 21st International Conference on Research and Education in Mechatronics, REM 2020 (2020). https://doi.org/10.1109/REM49740.2020.9313905

Mule, S., Hehenberger, P., Plateaux, R., Penas, O., Patalano, S., Vitolo, F.: An approach and an illustrative case study for a hybrid development process in mechatronic system design. Int. J. Prod. Lifecycle Manag. **13**(3) (2021). https://doi.org/10.1504/IJPLM.2021.118044

Munns, A.K., Bjeirmi, B.F.: The role of project management in achieving project success. Int. J. Proj. Manag. **14**(2) (1996). https://doi.org/10.1016/0263-7863(95)00057-7

Oehmen, J.: Projektmanagement. In: Pahl/Beitz Konstruktionslehre, pp. 877–879 (2021)

Oehmen, J., Olechowski, A., Robert Kenley, C., Ben-Daya, M.: Analysis of the effect of risk management practices on the performance of new product development programs. Technovation **34**(8) (2014). https://doi.org/10.1016/j.technovation.2013.12.005

PMI: A Guide to the Project Management Body of Knowledge Project Management Institute. Project Management Institute (2017a). (Preprint)

PMI: PMBOK Guide | Project Management Institute, PMBOK Guide (2017b)

Portman, H.: Review Standish Group – CHAOS 2020: Beyond Infinity, Henny Portman's Blog (2021)

Project Management Institute: PMO Frameworks Report | PMI Pulse of Profession. Project Management Institute (2013). (Preprint)

Radujković, M., Sjekavica, M.: Project management success factors. Procedia Eng. (2017). https://doi.org/10.1016/j.proeng.2017.08.048

Rehder, A., Souza, J.V., Marx, R., Salerno, M.S.: Emerging field or passing fashion? A case study of Agile-Stage-Gate model in innovation processes. Revista de Gestao (2023). https://doi.org/10.1108/REGE-08-2021-0149. (Preprint)

Roozenburg, N.F.M., Cross, N.G.: Models of the design process: integrating across the disciplines. Des. Stud. **12**(4) (1991). https://doi.org/10.1016/0142-694X(91)90034-T

Schmidt, T., Atzberger, A., Gerling, C., Schrof, J., Weiss, S., Paetzold, K.: Agile Development of Physical Products - An Empirical Study about Potentials, Transition and Applicability (2019)

Serrador, P., Pinto, J.K.: Does agile work? - A quantitative analysis of agile project success. Int. J. Proj. Manag. **33**(5) (2015). https://doi.org/10.1016/j.ijproman.2015.01.006

Shenhar, A.J., Dvir, D.: Toward a typological theory of project management. Res. Policy **25**(4) (1996). https://doi.org/10.1016/0048-7333(95)00877-2

Smith, R.P., Morrow, J.A.: Product development process modeling. Des. Stud. **20**(3) (1999). https://doi.org/10.1016/s0142-694x(98)00018-0

Sommer, A.F., Dukovska-Popovska, I., Steger-Jensen, K.: Barriers towards integrated product development - challenges from a holistic project management perspective. Int. J. Proj. Manag. **32**(6) (2014). https://doi.org/10.1016/j.ijproman.2013.10.013

Timinger, H., Schmidtner, M., Reiche, F.: A framework for the construction and tailoring of engineering development process models. IEEE Trans. Eng. Manag. (2022). https://doi.org/10.1109/TEM.2021.3132699. (Preprint)

Tiwana, A., Keil, M.: The one-minute risk assessment tool. Commun. ACM (2004). https://doi.org/10.1145/1029496.1029497

Turner, R., Ledwith, A.: Project management in small to medium-sized enterprises: fitting the practices to the needs of the firm to deliver benefit. J. Small Bus. Manag. **56**(3) (2018). https://doi.org/10.1111/jsbm.12265

Ulrich, K.T., Eppinger, S.D., Yang, M.C.: Product Design and Development, 7th edn. Mc-Graw-Hill Education (2020)

de Vasconcelos Gomes, L.A., Seixas Reis de Paula, R.A., Figueiredo Facin, A.L., Chagas Brasil, V., Sergio Salerno, M.: Design principles of hybrid approaches in new product development: a systematic literature review. R and D Manag. (2021). https://doi.org/10.1111/radm.12476. (Preprint)

VDI/VDE 2206: Development of mechatronic and cyber-physical systems (VDI/VDE 2206) (2021)

VDI2221: Systematic approach to the design of technical systems and products (VDI2221) (2019)

Vinekar, V., Slinkman, C.W., Nerur, S.: Can agile and traditional systems development approaches coexist? An ambidextrous view. Inf. Syst. Manag. **23**(3) (2006). https://doi.org/10.1201/1078.10580530/46108.23.3.20060601/93705.4

Wynn, D., Clarkson, J.: 'Chapter 1 Models of Designing', Design Process Improvement - A Review of Current Practice (2005). (Preprint)

Wynn, D.C., Clarkson, P.J.: Process models in design and development. Res. Eng. Des. **29**(2) (2018). https://doi.org/10.1007/s00163-017-0262-7

Wysocki, R.K.: Effective Project Management: Traditional, Agile, Extreme. Wiley, Hoboken (2014)

Zasa, F.P., Patrucco, A., Pellizzoni, E.: Managing the hybrid organization: how can agile and traditional project management coexist? Res. Technol. Manag. **64**(1) 2020). https://doi.org/10.1080/08956308.2021.1843331

Agile Transformation in Hardware Product Development - A Comparative Case Study

Lisa Siewert[1(✉)], Alexander Atzberger[2], Alexander Laukemann[3], and Kilian Gericke[1]

[1] University of Rostock, Rostock, Germany
lisa.siewert@uni-rostock.de
[2] Webasto SE, Stockdorf, Germany
[3] Hilti, Schaan, Liechtenstein

Abstract. Organizations increasingly adopt agile project management in hardware development to enhance flexibility and innovation, yet transitioning from traditional methods poses challenges like resistance, inadequate training, and resource constraints. This study examines two companies' eight-year transformations, emphasizing the role of change management in addressing these barriers. Key strategies include leadership commitment, tailored training, cross-functional teams and fostering collaboration. While limited by sample size, the findings provide actionable insights for aligning agile practices with strategic goals and driving sustainable success in dynamic industries.

1 Introduction

The German manager Rudolf von Benningsen-Foerder once stated that "standing still means going backwards" (Lederer 2016). In accordance with this slogan, organizations are continually striving to change and adapt their operations to today's fast-paced and highly competitive business environment. In order to remain competitive in this increasingly complex and dynamic environment, companies are forced to make significant investments in change (Burnes 2011). In the past years, companies increasingly turned to agile project management methodologies to enhance their flexibility, speed and ability to innovate. Originally developed for software development, agile methods have expanded into various sectors, including hardware product development, where traditional, linear project management approaches have long been dominant (Cooper 2011; Atzberger and Paetzold 2019). However, the transition to agile methodologies presents significant challenges, particularly in industries accustomed to well-established, rigid structures. Ovesen and Dowlen (2012) have summarized a broad set of challenges associated with the application of agile methods which were reviewed and updated seven years later by Atzberger and Paetzold (2019). The current challenges when applying agile project management in hardware development can be assigned to four main categories: Constraints of physicality, Mindset, Scaling and Team Distribution. Challenges allocated to the "Mindset" are proper education and training and the willingness to accept a new working style as well as engagement of top and middle management. Additionally, refusal and frustration increase if the capabilities, limitations and possible adaptions of agile methods are not communicated properly (Atzberger and Paetzold 2019). These challenges are also an integral part of research in change management.

K. Gericke et al. (Eds.): DSC 2022/2023, PTT, pp. 70–89, 2026.
https://doi.org/10.1007/978-3-032-16278-6_6

A change management model acts as a guiding framework, helping to navigate the complexities of transformation by outlining specific processes and steps to follow. It also highlights the various factors that influence change and identifies the key levers that can be used to achieve success in the change management process (Errida and Lotfi 2021). Change management plays a critical role in ensuring the successful implementation of agile practices within organizations. Effective change management strategies are essential to overcoming resistance, reshaping corporate culture, and aligning organizational processes with the principles of agility (Bader et al. 2024). The success of agile transformation hinges on how well the organization manages this change, making it essential to study the intersection of change management and agile project management, which is going to lead the following chapter with the consequential research questions:

RQ 1: What are key strategies that facilitate the implementation of agile project management in hardware product development?
RQ 2: How does organisational culture affect the agile transformation in hardware-focused companies?
RQ 3: What are common challenges and barriers faced by companies during the agile transformation process, and how can these be mitigated through effective change management?

2 Theoretical Background

2.1 Change Management

Change management is a structured approach that ensures organizational change is implemented purposefully and successfully, with minimal disruption. It involves preparing, supporting, and helping individuals, teams, and organizations in achieving the desired organizational change.

Change is essential for growth, yet it is often challenging because we are inherently drawn to routines and familiar patterns (Galli 2018). Studies show that in most organizations two out of three transformation initiatives fail (Sirkin, Keenan and Jackson 2005; Burnes 2011). Bader et al. (2024) presents in her research the failure factors that indicate a high probability for projects to fail. The factors were grouped into five categories: management and leadership, organization, operation, human resource management and external factors. The success of a transformation project hinges on sustained organizational motivation, derived from the necessity of the transformation and its acceptance within the organization (Lameijer et al. 2021). Human resources play a crucial role in this process, but resistance to change is a significant barrier as a primary cause of failure. Additionally, "lack of top management support" and "lack of training" are critical failure factors. These issues are consistent across various industries and organizational sizes, underscoring their widespread impact (Bader et al. 2024). These factors are interrelated; for instance, insufficient top management support can exacerbate resistance to change, which may be further fueled by inadequate training and communication. Ultimately, the organizational system must be flexible to proactively address obstacles, as unmanaged issues can lead to project failure (Belassi and Tukel 1996). Errida and Lotfi (2021) on the other hand display factors that enhance the success rate of transformation projects. One of the crucial factors for success is acceptance by all employees and stakeholders.

Therefore establishing a clear and shared vision aligned with organizational strategy is essential (Whelan-Berry and Somerville 2010; Smith et al. 2014). This vision should outline the future state, reasons for change, and expected outcomes (The Association of Change Management Professionals 2014). Additionally, organizations must assess their readiness for change at both individual and organizational levels. This includes preparing employees with the necessary skills and motivation as well as ensuring the organization has the necessary resources to support the transformation and also to keep it running (Project Management Institute 2013). In line with this, managing resistance is critical, as resistance is a major barrier to change (Errida and Lotfi 2021). Identifying and addressing the sources of resistance early on can significantly impact the success of change initiatives. Consistent and effective communication is essential for reducing resistance, creating change readiness, and gaining buy-in from stakeholders (The Association of Change Management Professionals 2014; Appelbaum et al. 2017). Communication should clearly convey the strategy, vision, and scope of the change project. Recognizing and rewarding short-term wins during the change process helps maintain momentum and keeps employees engaged and motivated (Mento, Jones and Dirndorfer 2002; Kotter 2007; Kanter 2011). It must also be mentioned that strong leadership and active sponsorship from top management are crucial for aligning stakeholders and ensuring the successful implementation and sustainability of change (Errida and Lotfi 2021).

Harold L. Sirkin, Perry Keenan and Alan Jackson (2005) agree that soft factors such as culture, leadership and motivation are important for success, but at the same time emphasise that these aspects alone are not sufficient to implement transformation projects. Instead, hard factors such as assessing the time required to complete it, the number of people needed to carry it out, and the financial results expected from the planned actions are even more important. In an extensive study, they found a consistent correlation between the outcomes (success or failure) of change programs and four hard factors: project duration, particularly the time between project reviews; performance integrity, or the skills of the project teams; the commitment of both senior management and the employees most affected by the change; and the additional effort required by employees to cope with the change. On this basis, they state that change projects fail to get off the ground when companies neglect the hard factors.

While identifying and implementing the critical success factors is essential for increasing the likelihood of successful change management, it is equally important to utilize a structured change management methodology. These methodologies provide a systematic approach that integrates the identified success factors into a cohesive process, ensuring that all aspects of change are addressed effectively (Parry et al. 2014; Rothwell, Stavros and Sullivan 2015; Stouten, Rousseau and De Cremer 2018). Parry et al. (2014) distinguish between processual and descriptive change management models. According to them, a processual model determines the steps for conducting and managing change. An example is Lewin's three-stage model (Lewin 1951), which is also considered the theoretical foundation of change management. Kurt Lewin's Change Management Model outlines three fundamental stages for managing organizational change: unfreezing, transition, and refreezing. In the "unfreezing" phase, the focus is on breaking down the current status quo, creating awareness and motivation for change. This is followed by the "transition" phase, where the organization implements the change, navigating through challenges and resistance. Finally, the "refreezing" phase ensures the change is solidified by establishing new behaviours, processes, and organizational culture, making the changes a permanent part of the system (Galli 2018; Errida and Lotfi 2021). Lewin emphasizes the importance of a well-communicated goal or motivation, because having the employees on board and having them adapting and incorporating new processes into their routine is crucial for success. Therefore it is important to have resources readily available if employees need training or want to make inquiries (Galli 2018). There are several other processual models, like Kotter's eight-step change model (Kotter 2007), Jick's ten-step change model (Jick 1991) or the seven-step Change Acceleration Process Model developed by the General Electric Company (Garvin 2000), which can all be traced back to the original three steps defined by Lewin.

Descriptive models on the other hand specify the main variables and factors that affect organizational performance and success. These models prioritize different aspects and offer varying perspectives on change (Errida and Lotfi 2021). The McKinsey 7-S Model, developed by Tom Peters, Richard Pascale, and Robert Waterman Jr., is a framework used to assess and implement organizational changes by examining seven interconnected elements: strategy, structure, systems, skills, staff, style, and shared values (Galli 2018). This model focuses on the internal factors of an organization, emphasizing how these elements must align for effective change. In contrast, the Burke-Litwin Model (Burke and Litwin 1992), the Congruence Model by Nadler and Tushman (Nadler and Tushman 1980), and other frameworks, such as those by Beckhard and Harris (Hamner, Beckhard and Harris 1978), and Carnall (Carnall 2010), extend the analysis to include external factors, cause-effect relationships, and the role of leadership, culture, and motivation in organizational change. These models often combine transactional and transformational factors to drive change and emphasize the importance of overcoming resistance through various strategic approaches and leadership competencies (Errida and Lotfi 2021).

2.2 Agile Project Management

Agile project management gained significant attention following the 2001 meeting of seventeen experts who captured the core values and principles in the *Agile Manifesto*. The manifesto aimed to realign focus from rigid governance towards more collaborative development, emphasizing individuals, working software, customer collaboration, and adaptability (Beck 2001; Sommer 2019). Agile development is characterized by flexibility, lightweight processes, and short-term iterations, allowing teams to actively involve users, validate needs, and prioritize requirements. Agile-managed projects are iterative, incremental, and self-organizing, adapting over time rather than being rigidly predefined (Boehm and Turner 2003). In mechatronic systems, agility enables teams to adapt quickly to changes in the development context (Albers et al. 2019).

Since the *Agile Manifesto*, agile project management has gained immense popularity, particularly in software development, due to its short development cycles, increased productivity, customer satisfaction, and higher product quality (Cho 2009; Gustavsson 2016). A study by Serrador and Pinto (2015) confirms that greater project success correlates with the extent of agile implementation. Agile principles have since been applied across various industries, including hardware product development, showing improvements in team collaboration, transparency, and flexibility (Albers et al. 2019; Schmidt et al. 2019). However, the application of agile project management in hardware contexts presents unique challenges, especially when implemented in a large-scale project (Zasa, Patrucco and Pellizzoni 2020). First of all, a successful implementation requires training, because agile project management can only be successful with talented individuals who favour many degrees of freedom and are willing to change their mindset and ways of working (Cho 2009; Sommer 2019). Other application challenges occur due to prototyping, especially realizing deliverables within a single iteration, and breaking down tasks into meaningful increments. External dependencies on a technical level, unsuitable corporate culture on a behavioural level, inappropriate corporate structure on an organizational level, add to the problem (Schmidt et al. 2019).

A variety of factors can be used to evaluate if agile project management is beneficial. Dybå and Dingsøyr (2008) present a list of internal and external context factors to distinguish agile and conventional development. Agile product development is most effective when applied in small development teams where the principles of continuous design improvement and testing are based on rapid feedback and change. Tools like the Hybrid Model Radar (Cocchi, Dosi and Vignoli 2023) and the *Agile Practice Guide* (PMI 2017) help managers assess when to apply agile versus plan-driven approaches, based on project complexity, innovation, and uncertainty.

The most commonly used agile practices are Scrum, Kanban and Design Thinking (Atzberger et al. 2020). Scrum is a framework designed to help teams work iteratively and incrementally. It breaks down projects into small, manageable pieces called sprints, which typically last two to four weeks. Each sprint begins with planning and ends with a review and retrospective to assess progress and make improvements. Scrum emphasizes daily stand-up meetings, where team members discuss progress, challenges, and next steps. The framework relies on key roles like the Product Owner, Scrum Master, and Development Team to ensure collaboration and focus on delivering high-quality, functional increments (Schwaber and Sutherland 2017). Kanban was first introduced by Taiichi Ohno as part of the Toyota Production System in the 1940s, aimed at improving manufacturing efficiency. It is a visual workflow management method designed to help teams optimize their work and ensure continuous delivery. It emphasizes visualizing tasks on a board, limiting work in progress (WIP), and improving efficiency by focusing on flow (Anderson 2011). Design Thinking is a human-centered approach to problem-solving that emphasizes empathy with users, creativity, and iteration. It typically follows five phases: empathize, define, ideate, prototype, and test. This methodology encourages teams to understand the needs of users deeply before defining problems and brainstorming potential solutions (Arnold 1956; Auernhammer and Roth 2023).

2.3 Agile Transformation

The term "agile transformation" is often ambiguous due to the scarcity of documented large-scale success stories. Due to this, the article at hand defines "agile transformation" as the process of shifting an organization's current state toward becoming an "agile organization." As there are only few definitions of adjacent terms such as an agile company (Dove 1999) or enterprise agility (Tsourveloudis and Valavanis 2002) available, an "agile organization", being the target state of an agile transformation, shall be defined as follows:

An agile organization is characterized by its adoption of the core principles outlined in the Agile Manifesto and its ability to respond swiftly to emerging changes. Agility needs to be understood as a capability which needs to be obtained, not a mere characteristic, hence it is not defined by the widespread use of specific frameworks or methodologies like Scrum or SAFe. Instead, agility should be measured by how quickly the organization can adapt to changing external and internal conditions, regardless of the tools or practices employed. The ability to respond flexibly and adaptively to various shifts is the true mark of agility (Quick 2024).

An important remark has to be added:

While scientific literature and real-world reports often portray organizations as lagging behind in the transformation process, this doesn't necessarily mean they strictly adhere to traditional "waterfall" methods as defined by Royce (1987). Instead, many employ Stage-Gate models, where parallel processes with domain-specific responsibilities are common, resembling simultaneous engineering efforts (Bers et al. 2014).

Agile's core tenets—iterative development, frequent customer feedback, and flexibility—are more difficult to implement in hardware or in mechatronics where prototyping and production have longer lead times (Albers et al. 2019). Nevertheless, several companies developing mechatronic products have already implemented agile methods in development projects that turn out to be quite successful (Kuberg and Komus 2015). When looking at the success factors for agility in an organization defined by Stelzmann et al. (2010) compared with the factors for successful change management by Errida and Lotfi (2021) there are several similarities. Both Stelzmann and Errida emphasize the importance of strong leadership. In agile project management, Stelzmann highlights the need for active and competent leadership to steer agile processes and decision-making. Similarly, Errida underlines the role of leadership in fostering organizational change, particularly through strong sponsorship from top management. Additionally, effective team collaboration and stakeholder engagement as well as adaptability and flexibility are important. While agile methodologies inherently value adaptability, the same factor is crucial for companies to effectively implement changes (Errida and Lotfi 2021).

In the context of transitioning to agile methodologies, change management is particularly challenging due to the need for a cultural shift, adoption of new practices, and the modification of existing workflows. The success of an agile transformation hinges on how well the organization manages this change, making it essential to study the intersection of change management and agile project management.

3 Study Design

For this study, two European companies were analysed using a comparative case study approach. Data was collected through semi-structured interviews with one employee from each organization. An initial open interview was conducted with both participants to discuss the general company context, similarities between the companies, key challenges and areas of focus. Based on this, an interview guide was developed and used in individual interviews with each participant. The guided interview structure was divided into five sections: (1) company background and the interviewee's role, (2) details of the transformation process, including its origins and future plans, (3) success stories from the transformation, (4) challenges encountered during the process, and (5) best practices and advice for others undergoing similar transformations. The interviews were supplemented with documents provided by the interviewees for additional context and insight. Following the evaluation of the interviews and documents, a joint debriefing session was held with both participants.

The first interviewee (CS1) is currently a strategic category manager but shared insights from their previous role as a global process manager, a position held for three years. In this former role, which functioned as an in-house consulting group across various business units, plants, and market organizations, the interviewee was responsible for integrating global processes into the organization and ensuring their uniform implementation, with agility being a key element. The second interviewee (CS2) has been an agile coach for three years, also in a global staff role. Their coaching focused on integrating agile project management into product development. Notably, the interviewee brought significant expertise to the role, having transitioned from an academic career, providing a strong foundation for guiding the adoption of agile methodologies.

4 Findings

4.1 Case Study 1: Organic Implementation Strategy

Company Profile

The first company is a multinational company with approximately 34 000 employees. It operates in the construction industry, specializing in products for construction, building maintenance, energy and manufacturing industries.

The company's agile transformation, initiated around 2015, was primarily driven by its employees, with support from top management but without direct involvement. The introduction of agile methodologies began with smaller projects where team members saw the benefits and advocated for broader adoption. The transformation was initially a grassroots movement, initiated by intrinsic motivated employees who advocated for the implementation of agile practices based on insights and learnings from other companies. This initiative was made possible by the company's innovative culture and the freedom to operate that it fosters.

The early stages were exploratory, with select teams experimenting with agile methods like Scrum and Kanban. This informal adoption revealed the potential of agile to drive efficiency, particularly in the areas of product development and project management. By 2017, the company began to realize the value of scaling these agile practices across more teams. This led to the formation of a network of early adopters who had successfully piloted agile projects, working together to train and guide others.

However, the process was not without challenges. While efforts were made to create a hybrid agile model to accommodate both traditional project management and agile approaches, there was no systematic company-wide rollout. Training programs were offered but were often on a voluntary basis and not widely promoted. Over time, certain business units became strong advocates for agile and provided their teams with dedicated agile coaches to support the transformation, particularly during the early, iterative phases of product development.

Overall, the company's transformation was organic and driven by bottom-up initiatives, with a mix of formal and informal methods to integrate agile practices across its operations.

Transformation Process

The company initiated its agile transformation around 2015, driven primarily by the employees rather than top management. The introduction of agile methodologies began with smaller projects where team members saw the benefits and advocated for broader adoption. Initially, the transformation was not a top-down initiative but rather a grassroots movement, with employees pushing for the implementation of agile practices based on external exchanges and learnings from other companies.

The early stages were exploratory, with selected teams experimenting with agile methods like Scrum and Kanban. This informal adoption revealed the potential of agile to drive efficiency, particularly in the areas of product development and project management. By 2017, the company began to realize the value of scaling these agile practices across more teams. This led to the formation of a network of early adopters who had successfully piloted agile projects, working together to train and guide others.

However, the process was not without challenges. While efforts were made to create a hybrid agile model to accommodate both traditional project management and agile approaches, there was no systematic company-wide rollout. Training programs were offered but were often on a voluntary basis and not widely promoted. Over time, certain business units became strong advocates for agility and provided their teams with dedicated agile coaches to support the transformation, particularly during the early, iterative phases of product development.

Overall, the company's transformation was organic and driven by bottom-up initiatives, with a mix of formal and informal methods to integrate agile practices across its operations.

Challenges and Solutions

The company faced several challenges including (1) limited opportunities for training and sufficient global support (2) resource constraints, (3) resistance to change, (4) different perspectives on needed profiles, (5) cultural shift and leadership and (6) tool adoption and consistency.

First to mention, initial training programs were voluntary and not actively promoted. Employees had to search for the agile training on their own, leading to low awareness. Additionally, the training empowered teams to independently adapt agile methodologies, requiring teams to adapt agile methodologies on their own. Second, offering training and external coaching for all teams across a large organization (with over 1,900 employees at the headquarters) posed significant resource challenges. Many project teams had to fund external agile coaches from their project budgets, which made it difficult to scale this support. Third, some of the more experienced developers were hesitant to adopt new tools and agile methods. This resistance was often tied to discomfort with both new processes and technologies like Jira, exacerbating acceptance issues across teams. But also management exhibited resistance due to the challenge of balancing the freedom of agile teams and the need to remain a structured oversight. Fourth, Human Resources is still in the process of enhancing their understanding of agile methodologies, such as Scrum and Backlog management, to provide better support for the transformation from HR perspective. As a result, HR often focused on certifications rather than genuine agile expertise with a dedicated direction, which is difficult to prove. As a result, the integration of agile methods into the broader organizational structure was slowed due to different orientations. Fifth, there was no strong top-down mandate or unified strategy to embed agile practices across the organization. While previous initiatives like the "Lean" methodology received strong support from top management, agile practices did not receive the same level of support which made it challenging to standardize agile practices across different departments. Also, agile transformation requires a cultural shift, especially around collaboration and role flexibility. The interviewee noted that the lack of clear role descriptions in many agile projects led to blurred responsibilities and team members adhering to rigid job descriptions, which caused friction. And sixth, introducing tools like Jira or implementing processes like Story Points or using 3D-printed prototypes were met with varying degrees of success. Some projects embraced these innovations, but their broad adoption was inconsistent, as teams were often unequipped or hesitant to embrace change.

Based on those challenges, solutions can be extracted that would improve future attempts to implement agile practices. For example, expanding the training program to include ongoing, hands-on support, such as assigning agile coaches to project teams, proved to be an effective approach in certain divisions. These coaches assisted teams through iterative phases of product development, helping to embed agile practices until the teams could operate independently. Additionally, redefining project management roles to incorporate agile principles was a crucial step in CS1. Agile coaching and resources like the Master Planning Board were introduced to streamline project management. This tool helped bridge the gap between traditional project management methods and agile practices, particularly in early project stages. Another profitable aspect would be to foster agile ambassadors. Encouraging the development strategies of agile ambassadors within the company would help drive the transformation. These individuals promote the agile mindset within their teams and departments, facilitating the adoption of new methods across the organization. To enhance HR's understanding of agile practices, greater emphasis should be placed on internal development and training. By aligning HR strategies with agile goals and integrating them into broader organizational priorities, the company could support its employees in acquiring the necessary agile skills and competencies. Another essential step towards agility is embedding agile principles into the company's long-term strategy and gaining buy-in from upper management. The experiences made in CS1 also show, that working with Original Equipment Manufacturers (OEMs) that were already using agile methods accelerate projects, demonstrating the potential for agile to speed up processes and reduce costs when properly implemented.

Outcome and Current Status of the Transformation Process

The company is still in the midst of its agile transformation. Some parts of the organization have fully embraced agile practices with positive outcomes. The use of Story Points, prototypes, and agile coaching has shown that when properly supported, agile can lead to faster, more cost-efficient project outcomes. However, the interview reveals that the transformation is far from complete.

The lack of a company-wide mandate and alignment with corporate strategy has hindered broader adoption. Agile is seen as a tool used by individual teams rather than a core part of the company's operational philosophy. This fragmentation limits the overall effectiveness of agile transformation, as it is not yet ingrained into the company's culture and processes.

While agile transformation with some dedicated projects has seen some success, it is not as crucial as the new corporate strategy. This strategy focuses on enhancing customer productivity, ensuring customer safety, and being the best partner for sustainability. With stronger leadership, clear goals, and a unified effort, the company can fully benefit from these priorities.

4.2 Case Study 2: Systematic Implementation Strategy

Company Profile

The second company (CS2) is a global automotive tier-1 supplier with over 15,000 employees, specializing in vehicle heating, cooling, and roof systems. The company primarily serves B2B clients while also offering customized products for B2C customers. Agile methodologies are central to CO2's project management, particularly in product development, where agility is used to manage innovation complexities and product customization. By adopting agility, CS2 has accelerated development cycles and improved interdepartmental collaboration, enabling quicker responses to market trends and customer demands.

Transformation Process

The transformation process of the company began in 2016 when it initiated its first experiments with agile project management methodologies. This initial phase was characterized by a bottom-up approach, driven by enthusiastic employees who sought to integrate agile practices into their teams. By 2020, the company recognized the potential of these methods and decided to scale them across the organization, supported by external consultants. This shift was a strategic decision made at the executive level, aiming to foster a more adaptive and efficient operational framework.

The transformation unfolded over two years, with the implementation of a tailored agile framework that engaged around 500 employees, up from just 100 initially. A cross-functional agile transformation team was established, comprising representatives from various departments, ensuring a diverse range of perspectives and expertise. Training was a crucial component of this process, with a comprehensive catalogue of courses designed to equip employees with the necessary skills for agile working.

Challenges and Solutions

The transformation process of the company has faced significant challenges and lessons, particularly as it attempted to integrate agile methodologies across multiple business units into its operations. The initiative began positively in 2016, gaining momentum through a bottom-up approach and the involvement of enthusiastic employees. The journey gained substantial momentum once the top management agreed to fund an entire scaled agile transformation in 2020. However, as the top management support was cancelled two years later due to changing economic situations, the transformation encountered severe roadblocks by the end of 2022. One of the most critical factors in this entire transformation was the shifting commitment from top management. A change in the executive board led to a re-evaluation of priorities, which stalled the ongoing efforts and created an environment of uncertainty among staff.

Besides that, significant impact, a number of additional challenges appeared along the journey. As employees grappled with fears about their career trajectories and the implications of new roles, the company struggled with silos and conflicting interests. The HR department failed to actively support and incorporate the agile initiative into their strategic goals which is essential as new ways of working and job descriptions need the approval and support of both HR managers as well as the works council. The absence of a strategic alignment between the agile objectives and HR's priorities hampered progress, revealing a disconnect between theoretical frameworks and practical application. This gap highlighted the need for stronger collaboration across departments, particularly in fostering leadership buy-in and engagement.

To address these challenges, the company invested heavily in communication, conducting regular meetings and workshops to keep employees informed and engaged. Open discussions were encouraged, providing a platform for employees to voice concerns and ask questions. This proactive approach aimed to build trust and reinforce the value of agile methodologies, particularly at the leadership level. Nevertheless, resistance persisted, often manifesting in pushback from those entrenched in traditional practices.

The establishment of cross-functional teams, comprising members from various departments, played a vital role in facilitating the transformation. These teams worked on developing a tailored agile framework, ensuring that different perspectives were considered and integrated. Training programs were also implemented to equip employees with the necessary skills, but participation was inconsistent, highlighting the need for more structured incentives.

Ultimately, while the company made strides in creating an agile culture, the success of these efforts hinged on sustained leadership support and continuous involvement from HR. The reliance on external coaches and advisors provided initial guidance, but the long-term sustainability of agile practices became questionable without a solid internal foundation. As the organization navigates the complexities of this transformation, it faces the ongoing challenge of maintaining momentum in the absence of active support and to avoid regressing into old habits that could undermine progress.

Outcome and Current Status of the Transformation Process

The agile transformation of the company shows progress, but significant challenges remain. By the tie of the interview the transformation process came to a halt and agile working was left optional for project teams. The interviewee emphasized that the original goal of better responding to customer requirements has been partially achieved, although lengthy product development processes persist. There is increasing pressure to reduce time-to-market and improve project quality, especially compared to competitors from Asia.

Clear prioritization and resource allocation would be necessary to advance the transformation. The interviewee highlights that strategically embedding agile methods and ensuring effective communication between leadership and employees are essential. In line with this, the interviewee stressed that strong leaders who understand and apply agile methods are crucial for success. The long-term goal is to create a flexible organization capable of swiftly adapting to market changes while delivering high-quality products promptly.

The ongoing challenge is that financial constraints often hinder the implementation of innovative projects. Therefore, a clear focus on core competencies and continuous support for employees through training and coaching are crucial to ensuring the success of the agile transformation.

4.3 Summary and Hard Facts

To summarise the interviews, Table 1 displays some of the key quantitative data of the companies being analysed. Despite differences in industry sector and company size, both organizations initiated their transformation processes in the same year and adopted a similar approach. This alignment provides a solid foundation for comparison, allowing us to examine how each organization has evolved over the past eight years. Such a comparison offers valuable insights into their respective prerequisites, suitability for agile methodologies, and strategic decisions made throughout their transformation journeys. This context sets up a meaningful discussion on the factors influencing their divergent paths and lessons learned along the way.

Additionally, the paths the companies took and the extent of the transformation are outlined in Fig. 1. The timetable shows that both companies had a very similar start with a continuous increase of applied agile methods on team level in the first three to four years. After CO 2 changed their strategy from bottom-up to top-down in 2020, the application was rapidly evolving at a broader scale, heading towards an agile organization. It can also be observed, that the graph is flattening after the sharp increase.

Table 1. Hard facts of the companies under review

	CS1	CS2
Industry	Construction	Automotive tier-1
No. of employees	34,000	15,000
Customer structure	B2C, direct sales	Mainly B2B (B2C)
Agile adoption approach	Bottom-up	Started Bottom-up, Top-down since 2020
Start of application of agile methods	2016	2016
Organizational structure	2 Business Areas (Electric Tools & Accessories / Fastening & Protection) 10 Business Units	2 Business Areas (Electrification & Roof Systems) 6 Business Units

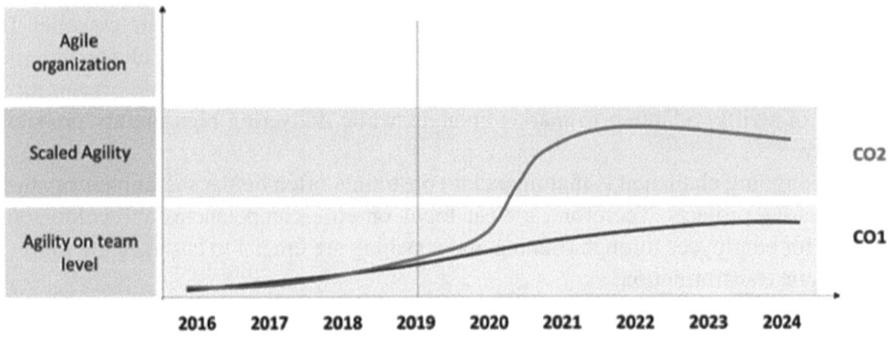

Fig. 1. Timeline of the transformation process

5 Discussion

5.1 Agile Suitability of CS1 and CS2

Implementing a new project management strategy as an overarching concept hinges on carefully integrated change management practices. Establishing the need for change and fostering a shared vision, as outlined in Kotter's model, is the initial step for aligning the organization's direction and motivating collective commitment (e.g. Kotter 2007). In agile project management, defining clear objectives from the outset remains as essential as in any other strategic shift. With well-defined goals, the organization can construct a roadmap that translates vision into actionable steps and measurable outcomes (Schock et al. 2024). Several tools can support decision-making on whether agile project management is suitable for a given organizational context, including the Agile Suitability Filter Tool (PMI 2017), which helps assess agile compatibility based on specific project factors.

The Agile Suitability Tool will serve to classify the two companies being studied, both of which began their transformation journeys in 2016. This classification offers a snapshot view, making the timing of the assessment crucial for a fair comparison. The selected reference point is 2019, three years post-transformation initiation, chosen for its balance: both companies shared comparable conditions at this stage, but this was prior to significant scaling efforts by CS2, after which the companies began to diverge in strategy and operations. This timing captures a critical juncture in the transformation process and offers a clearer basis for comparative insights (Fig. 1).

Figure 2 presents the Agile Suitability Tool analysis, comparing the readiness of the two companies for agile project management after an equal duration of transformation. The results indicate that CS1 demonstrates strong potential for agile methodologies due to high scores in experience, trust and access. However, factors such as buy-in, delivery and changes show a tilt toward a hybrid or predictive approach, suggesting that while CS1 is primed for agility, a fully agile model may not yet be optimal for every aspect of its operations.

In contrast, CS2 leans more consistently towards agile working across most criteria, though it shows a preference for hybrid methods in experience, trust and criticality. This profile suggests that CS2 is more uniformly positioned for agility but may benefit from a hybrid approach in areas requiring greater expertise and risk management. Overall, the chart provides insight into each company's distinct readiness for agile methodologies.

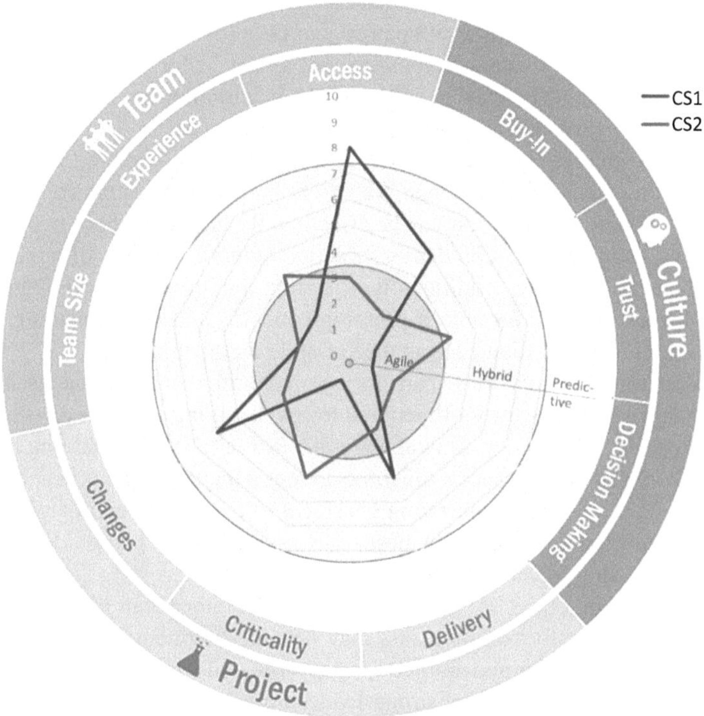

Fig. 2. Agile Suitability of the companies under review

5.2 Challenges and Risk Mitigation Strategies

Companies undergoing agile transformation often encounter a range of challenges that can impede their progress (Table 2). One primary barrier is limited training and awareness surrounding agile methodologies, which can lead to inconsistent understanding and implementation among employees. To address this, organizations should prioritize comprehensive and structured training programs that not only provide foundational knowledge but also ongoing support through mentorship and coaching. This approach is supported by current best practices, which emphasize the importance of experiential learning and the integration of agile principles into daily workflows.

One of the most common reasons why transformations fail can be retraced to limited top-management awareness. Especially with large-scale transformations, such as an agile transformation which has an impact on multiple departments across all levels, a strong awareness and commitment of the top-management is inevitable. Here, promoting agile ambassadors on both employee and board level is a significant success factor for an agile transformation. These individuals advocate for agile principles, facilitating knowledge sharing and fostering a culture of agility. Additionally, embedding agile methodologies into the company's long-term strategic objectives ensures alignment and sustained commitment.

Resource constraints frequently hinder the implementation of agile practices, especially in large organizations where external coaching and training must be funded from project budgets. To mitigate this, companies can establish dedicated budgets for agile initiatives, ensuring that adequate resources are allocated for training, coaching, and tool adoption. Leveraging internal expertise to conduct training sessions can also reduce costs while fostering a sense of ownership among employees.

Resistance to change, particularly from employees accustomed to traditional methods, poses another significant challenge. Research shows that engaging these individuals early in the transformation process can be effective. By including them in pilot projects or agile teams, organizations can demonstrate the tangible benefits of agile practices, thereby reducing resistance. Additionally, fostering a culture of psychological safety where employees feel comfortable expressing concerns and experimenting with new approaches can further facilitate acceptance of agile methodologies.

Siloed departments often create barriers to collaboration and knowledge sharing. To counteract this, establishing cross-functional teams that bring together diverse perspectives and expertise is essential. This aligns with current thinking that emphasizes the role of collaboration in agile success. Moreover, ensuring that agile practices and tools are standardized across the organization is critical to preventing confusion and inconsistency. Regular training sessions on these tools, coupled with clear guidelines for their use, can enhance adoption and integration.

Finally, if agile practices are perceived as optional rather than integral to the company's operations, transformation efforts may become fragmented. Research suggests that developing a cohesive and unified strategy for agile implementation is crucial. This strategy should be backed by strong leadership commitment and alignment with organizational goals. Regularly communicating the benefits of agile transformation and celebrating successes can reinforce the importance of these practices, helping to cultivate an agile mindset throughout the organization. By addressing these challenges with proactive strategies grounded in current best practices, companies can enhance their agile transformation efforts and drive sustainable change.

Table 2. Challenges and risk mitigation strategies

Challenge	Impact CS1	Impact CS2	Risk mitigation
Limited training options	○	○	• External support • In-house trainings • Self-learning offer (tailored learning paths for specific roles) • Onsite project coaching
Limited top management awareness	●	Beginning: ○ Later: ●	• Agile ambassadors • "Leading by example"
Resource constraints	◑	◑	• Dedicated budgets • Clear (new) role profiles • Supported transition from old to new roles
Resistance to change	○	◑	• Early engagement
Siloed departments	●	◑	• Cross-functional teams • Community of practices
Fragmented transformation efforts	◑	○	• Unified strategy • Culture journey

○: low, ◑: medium, ●: high

5.3 Key Strategies to Implement Agile Project Management

Organizational culture and leadership significantly influence the success of agile transformation in hardware-focused companies. Strong leadership commitment is critical; when top management actively supports agile initiatives, it fosters an environment that encourages change and innovation. However, shifts in executive priorities can hinder progress, underscoring the need for consistent leadership backing throughout the transformation process.

The implementation of agile project management in hardware product development is facilitated by several key strategies. A grassroots approach, where employees initiate the adoption of agile practices, often serves as a powerful catalyst for change. By starting with pilot projects, teams can experiment with methodologies like Scrum and Kanban, demonstrating tangible benefits and building confidence among participants. This incremental approach allows for learning and adaptation before scaling agile practices across the organization.

Another crucial strategy is the establishment of cross-functional teams that integrate diverse departmental perspectives. This diversity ensures that the agile framework is robust and tailored to the unique challenges of hardware development. Ongoing training and support are also vital; assigning agile coaches to teams provides the necessary guidance and mentorship, helping to embed agile practices effectively.

A culture that embraces collaboration and flexibility is essential for enabling agile practices. Employees must feel comfortable adapting to new roles and responsibilities, which may involve a departure from traditional job descriptions. Resistance to change, particularly from experienced employees accustomed to established methods, can pose challenges. Therefore, addressing this resistance through open communication and involving hesitant individuals in pilot projects can demonstrate the benefits of agile methodologies.

Additionally, aligning human resources with agile goals is vital for fostering an agile culture. HR should understand agile practices to support training and development initiatives effectively. Encouraging open dialogue between leadership and employees helps build trust, addresses concerns, and reinforces the value of agility within the organization. Overall, a supportive leadership presence and a culture conducive to collaboration and adaptability are crucial for successful agile transformation.

Ultimately, a combination of grassroots initiatives, cross-functional collaboration, continuous support, and strategic alignment and commitment from the top-management creates a solid foundation for successful agile implementation in hardware product development.

6 Conclusion

This chapter outlines the transformation process of two companies that started applying agile methods eight years ago. It emphasizes a structured approach for implementing agile project management, emphasizing the importance of change management to establish a shared vision and motivation. The readiness and specific requirements of the respective companies were analysed to underscore the need for tailored change management strategies.

The chapter highlights the key challenges companies face in agile transformations and proposes solutions to overcome them. Major obstacles include limited training, lack of top-management support, resource constraints, resistance to change, and departmental silos. Addressing these challenges requires structured training programs, active involvement from top management, adequate budgeting, cross-functional collaboration, and clear integration of agile practices into strategic objectives. Promoting agile ambassadors, fostering psychological safety, and creating unified strategies can further solidify the transformation process. By proactively tackling these areas, companies can enhance their agility and drive sustainable organizational change.

Successful agile transformation in mechatronic companies relies on strong leadership commitment, a supportive organizational culture, and a strategic approach to implementation. Consistent leadership backing throughout the transformation process is essential, while grassroots initiatives, such as pilot projects, help to introduce agile practices gradually. Establishing cross-functional teams and fostering an environment that values collaboration and adaptability further reinforce agile adoption. Addressing resistance to change, particularly from experienced employees, through open communication and pilot involvement also supports the transformation. Ultimately, a combined approach involving strategic alignment, employee empowerment, and continuous support creates a solid foundation for effective agile implementation in hardware development.

This chapter is limited by the small sample size. To provide more generalized insights and make more accurate conclusions, future research should expand the study to include a larger pool of companies and gather perspectives from diverse roles within each organization. This broader approach would enhance the validity of the findings and allow for a more comprehensive discussion of the research questions.

References

Albers, A., Heimicke, J., Müller, J., Spadinger, M.: Agility and its features in mechatronic system development: a systematic literature review. In: ISPIM Conference Proceedings (2019). (Preprint)

Anderson, D.J.: Kanban: Evolutionäres Change Management für IT-Organisationen (2011)

Appelbaum, S.H., Cameron, A., Ensink, F., Hazarika, J., Attir, R., Ezzedine, R., Shekhar, V.: Factors that impact the success of an organizational change: a case study analysis. Ind. Commer. Train. **49**(5) (2017). https://doi.org/10.1108/ICT-02-2017-0006.

Arnold, J.E.: Creativity in engineering. SAE Technical Papers (1956). https://doi.org/10.4271/560004

Atzberger, A., Nicklas, S.J., Schrof, J., Weiss, S., Paetzold, K.: Agile development of physical products: a study on the current state of industrial practice. Universität der Bundeswehr München, Neubiberg, Germany (2020).https://doi.org/10.18726/2020_5

Atzberger, A., Paetzold, K.: Current challenges of agile hardware development: what are still the pain points nowadays?. In: Proceedings of the International Conference on Engineering Design. ICED (2019). https://doi.org/10.1017/dsi.2019.227

Auernhammer, J., Roth, B.: What Is Design Thinking? (2023). https://doi.org/10.1007/978-3-031-36103-6_9

Bader, M., et al.: Why do process improvement projects fail in organizations? A review and future research agenda. Int. J. Lean Six Sigma **15**(3) (2024). https://doi.org/10.1108/IJLSS-07-2023-0126

Beck, K.: Manifesto for Agile Software Development (2001). https://agilemanifesto.org/

Belassi, W., Tukel, O.I.: A new framework for determining critical success/failure factors in projects. Int. J. Proj. Manag. **14**(3) (1996). https://doi.org/10.1016/0263-7863(95)00064-X

Bers, J.A., Dismukes, J.P., Mehserle, D., Rowe, C.: Extending the stage-gate model to radical innovation - the accelerated radical innovation model. J. Knowl. Econ. **5**(4) (2014). https://doi.org/10.1007/s13132-012-0131-6

Boehm, B.W., Turner, R.: Using risk to balance driven methods. PoLAR, 36, June 2003

Burke, W.W., Litwin, G.H.: A causal model of organizational performance and change. J. Manag. **18**(3) (1992). https://doi.org/10.1177/014920639201800306.

Burnes, B.: Introduction: why does change fail, and what can we do about it? J. Change Manag. (2011). https://doi.org/10.1080/14697017.2011.630507

Carnall, C.: Managing change in organizations. Int. J. Bus. Perform. Manag. (2010)

Cho, J.: A hybrid software development method for large-scale projects: rational unified process with scrum. Issues Inf. Syst., 340–348 (2009). https://doi.org/10.48009/2_iis_2009_340-348

Cocchi, N., Dosi, C., Vignoli, M.: Stage-gate hybridization beyond agile: conceptual review, synthesis, and research agenda. IEEE Trans. Eng. Manag. (2023). https://doi.org/10.1109/tem.2023.3282269.

Cooper, R.G.: Winning at new products: creating value through innovation (2011)

Dove, R.: Knowledge management, response ability, and the agile enterprise. J. Knowl. Manag. **3**(1) (1999). https://doi.org/10.1108/13673279910259367

Dybå, T., Dingsøyr, T.: Empirical studies of agile software development: a systematic review. Inf. Softw. Technol. (2008). https://doi.org/10.1016/j.infsof.2008.01.006

Errida, A., Lotfi, B.: The determinants of organizational change management success: literature review and case study. Int. J. Eng. Bus. Manag. **13** (2021). https://doi.org/10.1177/18479790211016273

Galli, B.J.: Change management models: a comparative analysis and concerns. IEEE Eng. Manag. Rev. **46**(3) (2018). https://doi.org/10.1109/EMR.2018.2866860

Garvin, D.A.: Learning in action: a guide to putting the learning organization to work. Harv. Bus. Sch. Press **79**(2) (2000). https://doi.org/10.1016/s0001-2092(06)60626-1

Gustavsson, T.: Benefits of agile project management in a non-software development context - a literature review. In: Project Management Development – Practice and Perspectives: Fifth International Scientific Conference on Project Management in the Baltic Countries, 14–15 April, Riga. University of Latvia, April 2016

Hamner, W.C., Beckhard, R., Harris, R.: Organizational transitions: managing complex change. Adm. Sci. Q. **23**(4) (1978). https://doi.org/10.2307/2392592

Sirkin, H.L., Keenan, P., Jackson, A.: The hard side of change management. Harv. Bus. Rev. (2005). https://doi.org/10.1109/emr.2014.6966953

Jick, T.D.: Implementing change: note. Harv. Bus. Sch. Cases (1991)

Kanter, R.M.: The change wheel: elements of systemic change and how to get change rolling. Harv. Bus. Sch. (2011)

Kotter, J.P.: Leading change: why transformation efforts fail. Harv. Bus. Rev. (2007). https://doi.org/10.1007/978-1-137-16511-4_7

Kuberg, M., Komus, A.: Studie "Status Quo Agile" – wie werden agile Methoden in der Praxis eingesetzt ? Projekt Magazin (13) (2015)

Lameijer, B.A., Antony, J., Chakraborty, A., Does, R.J.M.M., Garza-Reyes, J.A.: The role of organisational motivation and coordination in continuous improvement implementations: an empirical research of process improvement project success. Total Qual. Manag. Bus. Excell. **32**(13–14) (2021). https://doi.org/10.1080/14783363.2020.1757422

Lederer, D.: Chefsache Gesundheit II, Chefsache Gesundheit II (2016). https://doi.org/10.1007/978-3-658-06962-9

Lewin, K.: Field theory in Social Science: Selected Theoretical Papers (Edited by Dorwin Cartwright) (1951)

Mento, A., Jones, R., Dirndorfer, W.: A change management process: grounded in both theory and practice. J. Change Manag. **3**(1) (2002). https://doi.org/10.1080/714042520

Nadler, D.A., Tushman, M.L.: A model for diagnosing organizational behaviour. Organ. Dyn. **9**(2) (1980). https://doi.org/10.1016/0090-2616(80)90039-X

Ovesen, N., Dowlen, C.: The challenges of becoming agile - experiences from new product development in industry and design education. In: Proceedings of the 14th International Conference on Engineering and Product Design Education: Design Education for Future Wellbeing, EPDE 2012 (2012)

Parry, W., Kirsch, C., Carey, P., Shaw, D.: Empirical development of a model of performance drivers in organizational change projects. J. Change Manag. **14**(1) (2014). https://doi.org/10.1080/14697017.2012.745894

PMI: Agile Practice Guide, Project Management Institute (2017)

Project Management Institute: PMO Frameworks Report | PMI Pulse of Profession. Project Management Institute (2013)

Quick, L.: Agility vs Agile: What are the differences and how are they similar? upGrad Knowledge (2024). https://www.knowledgehut.com/blog/agile/agility-vs-agile

Rothwell, W.J., Stavros, J.M., Sullivan, R.L.: Organization Development, Transformation, and Change. In: Practicing Organization Development: Leading Transformational Change, 4th edn. (2015). https://doi.org/10.1002/9781119176626.ch1
Royce, W.W.: Managing the development of large software systems: concepts and techniques. In: ICSE 1987 Proceedings of the 9th International Conference on Software Engineering (1987)
Schmidt, T., Atzberger, A., Gerling, C., Schrof, J., Weiss, S., Paetzold, K.: Agile development of physical products - an empirical study about potentials, transition and applicability (2019)
Schock, M., Muller, J., Batora, M., Albers, A.: A comprehensive approach for the strategic determination of individual target levels of agility in product development. In: Proceedings of NordDesign 2024, pp. 711–719. The Design Society (2024). https://doi.org/10.35199/NORDDESIGN2024.76
Schwaber, K., Sutherland, J.: The Scrum Guide: The Definitive The Rules of the Game. Scrum.Org and ScrumInc, November 2017. (Preprint)
Serrador, P., Pinto, J.K.: Does agile work? - a quantitative analysis of agile project success. Int. J. Proj. Manag. **33**(5) (2015). https://doi.org/10.1016/j.ijproman.2015.01.006
Smith, R., King, D., Sidhu, R., Skelsey, D.: The Effective Change Manager's Handbook. Essential Guidance to the Change Management Body of Knowledge, Animal Genetics (2014)
Sommer, A.F.: Agile transformation at LEGO group: implementing agile methods in multiple departments changed not only processes but also employees' behavior and mindset. Res. Technol. Manag. **62**(5) (2019). https://doi.org/10.1080/08956308.2019.1638486
Stelzmann, E., Kreiner, C., Spork, G., Messnarz, R., Koenig, F.: Agility meets systems engineering: a catalogue of success factors from industry practice. Commun. Comput. Inf. Sci. (2010). https://doi.org/10.1007/978-3-642-15666-3_22
Stouten, J., Rousseau, D.M., De Cremer, D.: Successful organizational change: Integrating the management practice and scholarly literatures. Acad. Manag. Ann. **12**(2) (2018). https://doi.org/10.5465/annals.2016.0095
The Association of Change Management Professionals: Standard for Change Management. The Association of Change Management Professionals (2014). (Preprint)
Tsourveloudis, N.C., Valavanis, K.P.: On the measurement of enterprise agility. J. Intell. Robot. Syst. Theory Appl. **33**(3) (2002). https://doi.org/10.1023/A:1015096909316
Whelan-Berry, K.S., Somerville, K.A.: Linking change drivers and the organizational change process: a review and synthesis. J. Change Manag. **10**(2) (2010). https://doi.org/10.1080/14697011003795651
Zasa, F.P., Patrucco, A., Pellizzoni, E.: Managing the hybrid organization: how can agile and traditional project management coexist? Res. Technol. Manag. **64**(1) (2020). https://doi.org/10.1080/08956308.2021.1843331

Applying Scrum to Ship Design

Junheng Zhang(✉) and Florian Sprenger

University of Rostock, Rostock, Germany
junheng.zhang@uni-rostock.de

Abstract. Recent environmental regulations in shipping and general post-pandemic trends in the working environment are introducing new circumstances for ship design. The newly implemented regulations by the International Maritime Organization (IMO) towards a massively reduced greenhouse gas (GHG) emission for global shipping will affect the ships significantly. To obey these regulations, existing and newly built ships will utilize new energy sources as well as energy-saving technologies (EST) in combination. Meanwhile, as an outcome of the COVID-19 pandemic crisis, shipyards and design offices need to adopt remote working solutions demanded by their employees. With all consequences derived from these effects, the ship design is facing an entire set of uncertainties and obstacles, challenging the way conventional ship design is governed. This article investigates the influences of each factor on the traditional ship design methodology commonly utilized at most shipyards and design offices. By breaking down the previously mentioned factors, the influence is qualitatively assessed. In addition, a design approach following the Scrum methodology is proposed to better cope with the ship design with many uncertain factors.

The finding concludes that not any single factor but the bundle of multiple components will require ship design to alter their conventional approach. The lack of experience with new energy resources and their potential interaction with ESDs can lead to an inefficient design process. Meanwhile, the remote working tendency in the post-pandemic era will require the ship design to move away from its traditional structure and adjust its governing accordingly to the new situation. The conventional ship design method following the design spiral will remain applicable and, hence, can serve as a foundation for the Scrum ship design. However, The organizational aspects must be adjusted according to remote working models. A suitable workflow of Scrum ship design is presented.

1 Introduction

The maritime sector is subject to swift changes. Tightened environmental requirements require new (as well as old but back in focus) propulsion systems and "addon"-devices to contribute to the ship's overall efficiency (Lappalainen et al. 2020; Hammer and Eide 2021). At the same time, two years of the global pandemic situation changed the perception of how work can be arranged for people in the design departments (Angle and Brackley 2022). Under this context, the shipbuilding industry is facing uncertainties throughout several fields, whether it is the ship as the product itself or the ship designer. Since the conventional ship design mostly relies on experience and therefore lacks the

K. Gericke et al. (Eds.): DSC 2022/2023, PTT, pp. 90–112, 2026.
https://doi.org/10.1007/978-3-032-16278-6_7

information foundation for the upcoming changes, a new ship design method follows the agile principles, more specifically the scrum method, is introduced to better cope with the uncertainties.

The global industry is facing accelerated efforts across all sectors, moving towards net-zero targets. Solutions utilizing renewable energy resources have been identified as a crucial factor in the fight against climate change in the transport sector. The European Union, e.g., aims to increase hydrogen production to 40 times by 2040 and use it for heavy-duty transport (Angle and Brackley 2022). Logically, the shipbuilding industry is no exception. The International Maritime Organization (IMO) introduced regulations in 2011 to reduce ship particle emissions (International Maritime Organisation 2011). These regulations will affect the fleet of all IMO-regulated countries and require ship operators to equip their ships with new propulsion systems along with energy-saving technologies (EST) to improve overall operation efficiency. The integration of these systems marks a change in the product ship itself as some of them significantly differ from long-established components. This requires ship designers to overthink their established methodology as, in most cases, no previous knowledge is available to rely on (Wanf et al. 2020).

Two years of the global pandemic have reshaped the conventional workplace and introduced new possibilities for workers and engineers. Flexible working models and locations enforced by COVID-19 led to many people rethinking how they want to work. These effects, alongside the increasing demand for specialists and the decrease in upcoming human resources, have led to a surge of the power leverage from employers toward employees (Angle and Brackley 2022). Shipyards and design offices are starting to adopt the concept of remote working into their job offers. The *Meyer Werft* has enabled their engineers to work from home (Meyer Werft Presse 2021).

The combined effect brings a set of uncertainties into the ship design. Meanwhile, software development utilizes engineering methods to successfully cope with uncertainties. Same as for ships, software programs are mostly uniquely designed and therefore considered as one-of-a-kind. Software consists of different components inter-affecting each other, similar to ships. In addition, the software environment is deemed fast-paced and subject to frequent changes (Errath et al. 2004). Agile methods aim to provide a flexible scope and efficiently handle uncertainties in a project. By embracing the uncertainty, agile has been successful in the field of Software design and is meanwhile introduced into different fields of engineering outside of software Does (Pedro Serrador and Jeffrey K. Pinto 2015). The most prominent method widely used in Software Engineering is Scrum, which is part of the agile framework (Sutherland 2012).

Under these circumstances, a ship design method based on the agile scrum framework is proposed in this article. First, the impact of the tightened environmental regulations on the prevalent ship design methodology shall be investigated. Followed by an overview of the effects of the remote working conditions. Based on these, the scrum ship design method is proposed.

2 Conventional Ship Design

A ship is usually equipped with a high number of components. A modern way to approach ship design is system-based and considers a ship as a complex system integrating task (Vossen et al. 2013; Papanikolaou 2014), involving suppliers of subsystems and components from other sectors. In this context, integration means not the sole physical integration but also, even more importantly, the warranty of smooth interactions between subsystems and components for an efficient ship operation. As an Engineering-to-Order (ETO) product and in contrast to Assemble-to-Order mass products like cars, ships are, in general, uniquely or at least for small numbers specified and subsequently also designed and produced (Ziółkowski and Dyl 2020).

Design Spiral

The conventional design methodology follows the first proposed way by Evans (Evans 1959). It is characterized by a sequential and cyclical approach to various ship design aspects and, hence, often depicted as a spiral (Fig. 1).

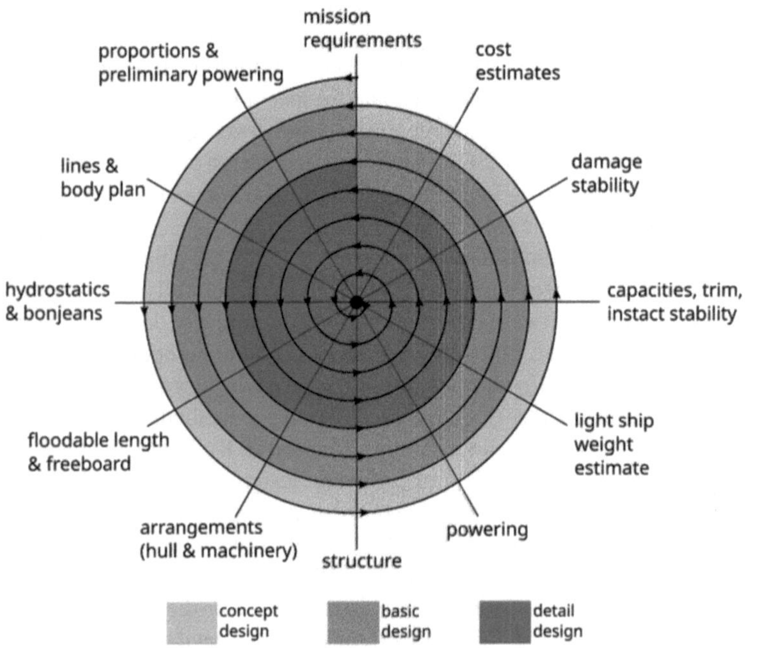

Fig. 1. Ship Design Spiral

The procedure is usually decomposed into four phases and typically follows an iterative approach. Those four phases target the design of the same technical aspects and components in a ship but vary in the level of detail. Each run towards the spiral center generates additional ship information regarding each aspect which is, in return, used to further specify other dependent aspects. The four phases are concept, preliminary, basic, and detail design, varying in the amount and detail of available information for the design process (Papanikolaou 2014).

The Concept Design marks the first phase and translates the mission requirements into basic ship characteristics, e.g., ship length, beam, draft, and block coefficient. Papanikolaou compares this phase equivalent to a feasibility study, which can be done within one day with the support of modern computer technology (Papanikolaou 2014).

The Preliminary Design addresses the aspects of Concept Design more refined. The outcome is the accurate ship main characteristics, though not fixed. This lies the foundation for the shipbuilding contract between the owner and the shipbuilder. This phase takes up to 15 times the concept phase duration (Papanikolaou 2014).

The third phase, Basic Design, investigates specific ship parts according to the mission requirements and determines accurate characteristics. The output of this phase includes naval architectural drawings and technical specifications of essential ship parts. At this point, information directly affecting the supply chain is formed, such as a detailed description of the hull shape and structural design (Papanikolaou 2014).

The last phase deals with the detailed design of the ship regarding its structural elements and concrete-specified equipment, such as machinery and other outfitting. The information generated here is typically directed to the production department of the shipyards and also to external suppliers for equipment and outfitting (Papanikolaou 2014).

For a long period of time, technological progress within the ship was more of an evolution rather than a revolution dominated by minor developments and refinements of prevalent technologies. New ship designs used to rely on existing vessel configurations and adjusted to up-to-date solutions (Ziółkowski and Dyl 2020). This approach was subject to continuous improvement and modification during the time in order to cope with different requirements depending on the ship type (Nowacki 2016). However, the overall main approach still relies on trial-and-error experiences, and existing ship designs are based on previously acquired knowledge in this field (Papanikolaou 2014). Nowadays, utilized ship design strategies are still emerging from Evans Design Spiral (Erikstad and Lagemann 2022). Thus, they inherit its characteristics.

Decarbonization of Shipping

The Decarbonization of global shipping is pushed with effort by the International Maritime Organization (IMO). The IMO operates under the United Nations and is responsible, among others, for the prevention of environmental pollution caused by global shipping. With almost 90% participating countries in the world, it virtually represents the entire globe (International Maritime Organisation 2023).

Energy Efficiency Design Index (EEDI).

IMO has introduced regulations to reduce CO2 emissions in the global shipping sector; one example is the Energy Efficiency Design Index (EEDI) (International Maritime Organisation 2020). It provides an approach moving towards the reduction of greenhouse gas emissions in the maritime shipping sector by up to 40% until 2030 based on the baseline rate from the year 2008. In general, EEDI puts the CO2-Emission in relation to the transport work $[\frac{g{\cdot}CO_2}{t.nm}]$ (Tokuslu 2020). In four phases starting with the year 2013, each new ship or ships with major conversion with a size of 400 gross tonnage and above must meet the EEDI requirements of the maximum allowed CO2 emissions (EEDIreq) for a certain transport work (Fig. 2) (International Maritime Organisation 2011). A similar regulation for existing ships, the Energy Efficiency Existing Ship Index (EEXI), is introduced as well (DNV 2022).

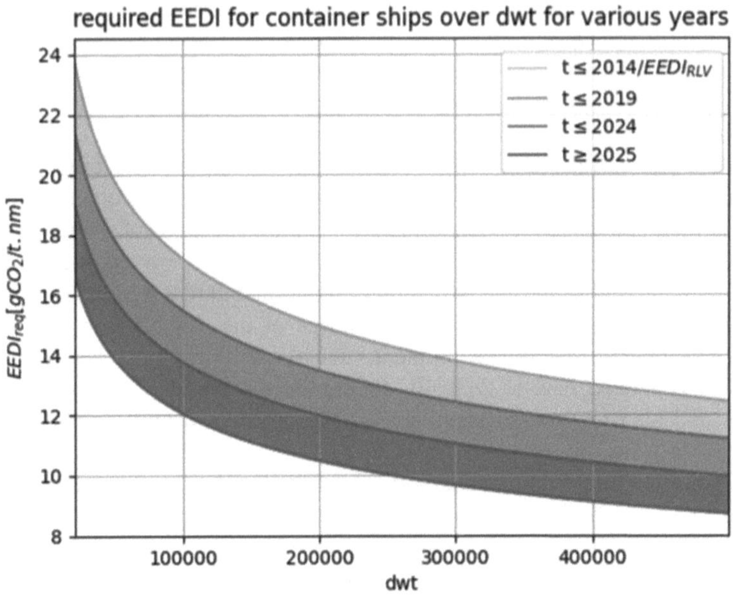

Fig. 2. . EEDI value for containership

Ship EEDI

The ship EEDI is calculated based on a single ship's components (International Maritime Organisation 2018). The main engine and auxiliary engines are taken into account as well as shaft generators, motor emissions, and any innovative technologies contributing to the efficiency. Measures targeting to reduce the ship EEDI will affect at least one of these components. The current focus lies on innovations for efficiency and alternative fuel options (Ovrum and Longva 2022; Ramboll, 2022; Liu and Ang 2021).

Innovations are intended to decrease the ship EEDI. It takes into account innovative energy-efficient technologies supporting the main engine to reduce potential CO_2 emissions (International Maritime Organisation 2018). Varying the fuel is a valid option to reduce the ship's EEDI. However, alternative fuels have different requirements that come with new technologies in shipbuilding. These technologies present new challenges for ship design and shipbuilding that need to be considered. This will be discussed later in this section.

There are several innovative technologies available whose implementation requires additional design effort. They are classified into three categories and further divided by the Availability (Table 1) (Germanischer Lloyd 2013). Category 1 contains technologies that will contribute to the propulsion power and reduce the required main engine propulsion power. Category 2 consists of technologies generating electrical power contributing to the energy generation on board. Category 3 can be deemed as the optimization of prevalent ship components. Therefore, it is already seamlessly accounted for in the common ship design process. Categories 1 and 2 consist of physical parts and components requiring dedicated treatment when it comes to the technical configuration and integration into ship design.

Table 1. Examples of innovative technologies decreasing EEDI (Germanischer Lloyd 2013)

Category	Availability	Example Technologies
1	all time	hull air lubricant system
	condition dependent	wind assistance
2	all time	waste heat recovery system
	condition dependent	photovoltaic cells
3	all time	hull optimization

The decision and selection of which GHG-saving technology shall be applied and respectively utilized for a ship depends on various aspects besides the integration effort and investment cost. The saving potential is also a crucial facet, and it varies among the currently available technologies and measurements (Table 2).

Table 2. Saving Potential of various ship technologies on GHG emissions (Khan et al. 2021)

Technology	Saving Potential [%]
wind assisted propulsion system	30
hull shape optimization	15
air lubrication	13
materials	10
improved propeller design	10
bulbous bow	7
heat recovery	6
hull surface treatment	5

Among the options, wind-assisted propulsion offers the most saving potential, given an optimal operation condition. The integration can require more effort, and therefore, design processes are affected more. The Flettner Rotor (FR), for example, contributes to the propulsion by utilizing wind energy through the Magnus-effect (Pearson 2014). Its integration requires additional structural support for the deck (Scandlines 2020), and its operation comes with several side effects, such as noise, vibrations, and additional heeling forces from the side of the ship (Pearson 2014). These need to be addressed by a more dedicated design to this use case.

With respect to the tightened environmental regulations and growing experiences with these technologies, one can assume an increasing role of wind-assisted propulsion systems in the foreseeable future.

New Ship Energy Source

Besides innovations having an assisting and/or supporting characters of main propulsion, research and development in shipbuilding are also focusing on possibilities to reduce emissions of main and auxiliary engines by using alternative energy sources to decrease the ship EEDI directly (e.g., the conversion between fuel consumption and the CO_2 emission of utilized fuel). Technologies using alternative fuels and energy sources are often considered as more promising options to reduce emissions. Fuel cells are integrated into the ship power system to reduce fuel consumption. The number of LNG-fueled ships is increasing and new builds are listed using LPG, batteries, hydrogen, methanol, or ammonia fuels.

Analyzing all realistically available technologies in the foreseeable future to decrease shipping GHG emissions, alternative fuels and energy sources are considered the most potential and only solution capable of achieving a 100% reduction (Table 3).

Table 3. Technologies and Potentials of Green House Gas Emission Reduction (Ovrum and Longva 2022)

Technology	Examples	Savings (up to)
Fuels and Energy	LNG, LPG, Ammonia, Methanol, Hydrogen, Harvesting from surrounding	100%
After Treatment Measure	Carbon Capture and Storage	30%
Logistics and Digitalization	Speed reduction, Routing	20%
Hydrodynamics	Hull Optimization, Air Lubrication	15%
Machinery	Machinery Improvements, Battery Hybridization,	20%

Obviously, not every single measure in the list applies to every ship type. Parameters regarding operation location and freight type are limiting factors. Ocean-going vessels with a global route rely on large energy storage. Operation profiles are less volatile but demand a steady power supply over a long time frame. Current realistic solutions offer mostly LNG and LPG, taking into account global and commercial scalability, while methanol technologies are on the verge of being widely introduced and already utilized in a few commercial applications (Hammer and Eide 2021). *MAERSK* expected the first delivery of their, in total, 19 ordered ocean-going cargo ships capable of running with methanol in 2023 (Serpa et al. 2023), which is postponed to early 2024 (MAERSK Press 2023). Ammonia is also identified as a promising solution but requires a longer time to establish since it comes with additional storage requirements and will demand new bunkering and handling technologies (Serpa et al. 2023).

For short-distance vessels, options are more variable. Operation profiles for these ship types are typically changeful in power demand and, therefore, suitable for solely electric or hybrid-electric energy and propulsion systems. In this context, short-sea shipping can serve as a field to gain more knowledge and experience with new technologies to mature (Hammer and Eide 2021). As most of the short-distance vessels are designed and manufactured in Europe, this offers a chance for European shipyards to acquire a competitive advantage in this field of shipbuilding (VSM AG 2021).

The selection of alternative fuels depends on, among others, operation profile and location. Aside from safety regulations and the technical readiness level of the utilized systems, the fuel energy density plays a key role in the consideration. When it comes to the transport task, ships are, in most cases, either volume- or mass-critical regarding cargo capacity. Hence, fuels with a high energy density per volume or/and mass are considered advantageous, as they occupy less potential cargo space. Current utilized and potential fuels vary in energy density (Fig. 3). The entire group of alternative fuels exhibits less specific energy density (volumetric and gravimetric) and, therefore, requests more space and buoyancy. Combined with the high price of alternatives compared to conventional fuels (Hammer and Eide 2021), this will inevitably lead to design adjustments.

Ship design must adopt new technologies in order to deliver competitive ships and meanwhile guarantee required regulations. Depending on the chosen fuel, crucial design implications, such as fuel storage, power plant, and safety requirements, need to be accounted for.

Consequently, it is reasonable to assume that not one single measure but the application of multiple measures on upcoming ships will be required to achieve the reduction goals for global GHG emissions. This implies changes in separate aspects of the ship design spiral, which, again, inter-influences, piling up in the demand of shipyards to rethink their design process entirely. However, all aspects, whether new technology or not, fit into the current design spiral. The underlying procedures might differ in detail, but the general approach follows the same pattern.

To sum up, from the technological perspective, all stated significance in this section fits into the design spiral since the functional aspect is still addressed. Hence, the design spiral as the given structure for the ship design remains an applicable approach, but the procedure of the design sequence and its planning is more influenced due to the lack of experience with the upcoming new technologies. A governing framework intended to cope with uncertainties can be beneficial under these circumstances.

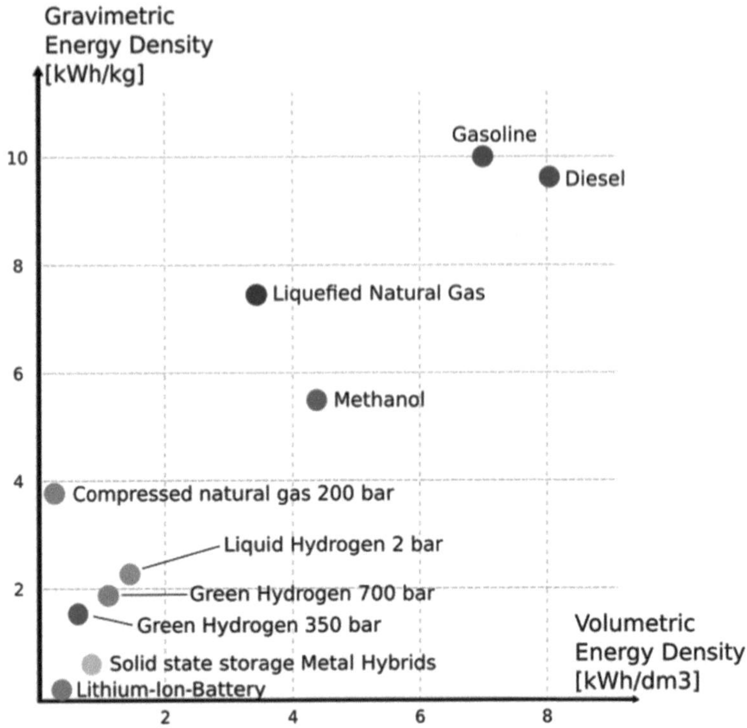

Fig. 3. Energy Density of Potential Ship Fuels 2021 (Voss 2020)

3 Remote Working Conventional Ship Design

Remote working gained attention over the recent years during the COVID-19 pandemic but is actually not a new topic. A definition of 'remote office work' delivered in 1983 describes it as organizational work that is conducted outside of the "organizational confines of space and time" (Olson 1983). In this context, (McTaggart and McLaughlin 2020) describe it as the possibility for employees to work at home as a substitution for attendance at the normal workplace on a regular basis. 'Home' can be replaced here with any arbitrary location suitable for work. For this definition, 'possibility' can be decomposed into two segments, which are pre-conditions for remote work. The first one is the technological advancements that enable remote working in general. This covers actual physical objects required for remote working, such as computer hardware and software. The second one is the organizational arrangements that allow their employees to not be physically present at the workplace. This should be accompanied by an organizational mindset toward remote working (McTaggart and McLaughlin 2020).

Acceptance of Remote Working

Nowadays, after two years under global pandemic circumstances, the working world has adopted a more flexible understanding of how work can be accomplished. The possibility of remote working will play a crucial role. A non-panel restricted survey conducted by

Fraunhofer Institute for Applied Information Technology reveals the following picture: Among over 3000 employees, over 75% reported working remotely on a daily basis and another 20% at least one time per week. A vast majority of over 60% are pleased to have this opportunity (Prinz 2021). And they are keen to keep it that way: *Deloitte* reported that 74% would choose to work at least partly from home, 49% even with more than half of their working time. This survey was carried out with almost 1700 employees in Switzerland (Deloitte 2021). 65% of the workers in the US are willing to waive five percent of their regular salary to receive full-time remote working opportunity in return, according to the Trend Report 2022 of *the SustainAbility Institute* (Angle and Brackley 2022).

On the employer side, another recent survey from 2022 among 530 companies conducted by KPMG finds that almost 90% have already or are in the process of establishing remote working standards (Burrows et al. 2021).

The working environment will inevitability become more flexible. Remote working options are more or less mandatory for companies if they want to stay on top of the competition for future talents (Fig. 4). Among the key drivers, employee-related reasons, such as following the request of their employees (25%), maintaining a good reputation (Employer branding) (18%) as well as talent shortage (12%), marks a major stake (Burrows et al. 2021).

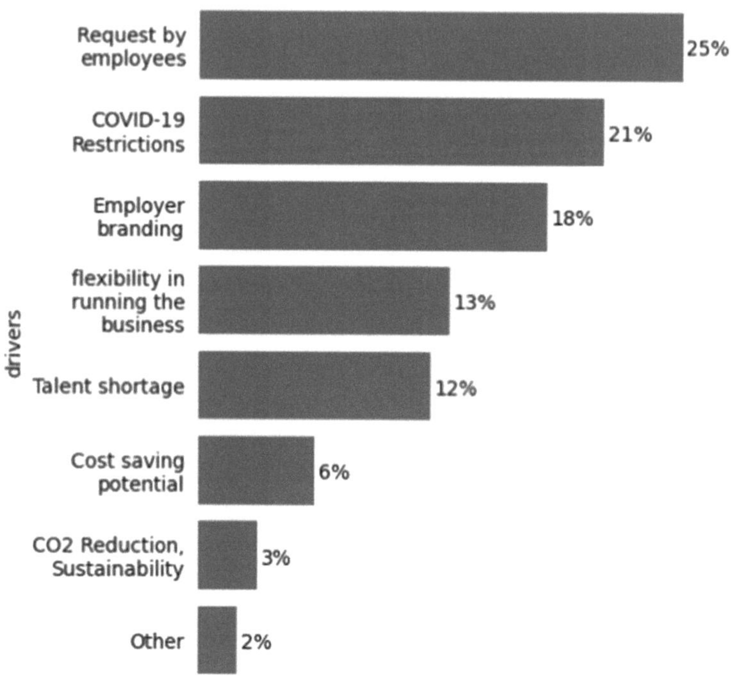

Fig. 4. Driver for company movement towards remote working (Burrows et al. 2021)

On the employee side, findings confirm a high level of contentedness among employees in the new situation with frequent remote working opportunities. Throughout all industry sectors, a vast majority of over 75% appreciate the flexible working location option. However, in fields relying on strong and frequent communication, and this applies to ship design with the involvement of many stakeholders on top, these tendencies are less strong (Prinz 2021). The dominant communication tool for information exchange is still E-Mail. Compared to conventional office work, the quality of team collaboration is reduced in general. Aspects regarding inter-support and social interactions are more affected, and meanwhile, information exchange in the working context is considered slightly worse compared to presence work (Prinz 2021). Here, room for improvement is identified.

Encouraging Communication

Investigations on the effect of remote working on information-related workers at *Microsoft* in the year 2020 have shown a general increase in information exchange. Whether regarding meetings and calls (Fig. 5) or messaging channels (Fig. 6), a significant increase is observed on the unplanned information exchange side when switching to remote working in the middle of February, whereas E-mail and scheduled meetings remain mostly constant (Yang et al. 2022). This indicates that remote working is only plannable to a certain extent. Spontaneous communication, whether work-related or not, is essential and, therefore, should be promoted by companies.

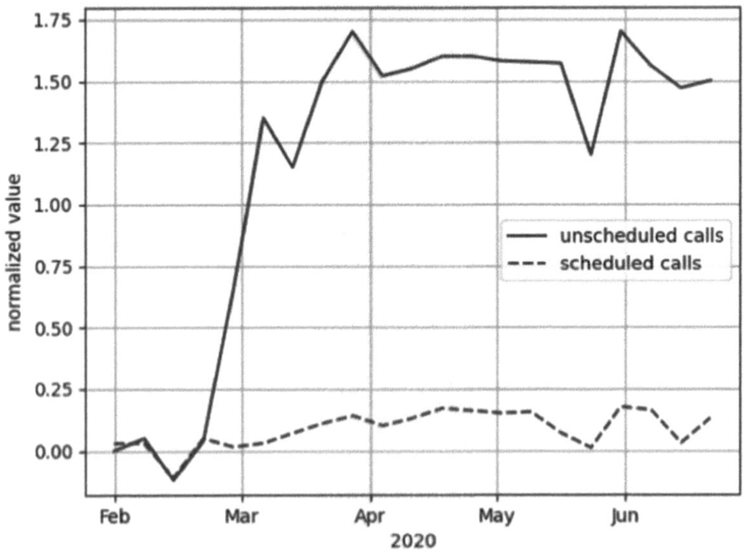

Fig. 5. Normalized numbers of scheduled calls vs. unscheduled calls in 2020 Feb–Jun at Microsoft (Yang et al. 2022)

Communication on a daily basis should not rely solely on E-Mail. Instead, a tool of a lower threshold is preferred to increase the tendency of designers to go into communication with each other, disregarding the topic being more small-talk related or professional.

As the lack of social interactions has been pointed out as the most negative impact for workers from remote sites during the pandemic situation, it is recommended that employees be motivated to go into conversation rather than work on their own. Therefore, taking any measures to reduce hesitation is one determining factor for companies.

Project Management under Remote Working Environment

Remote Working can make it more difficult to track the overall progress of a project. In ship design, the increase of information over the iterations will inevitably require better information management to monitor the actual progress. Hence, it is recommended to reduce the tasks set for designers into smaller portions where the task scope is not driven by content but by a fixed time frame.

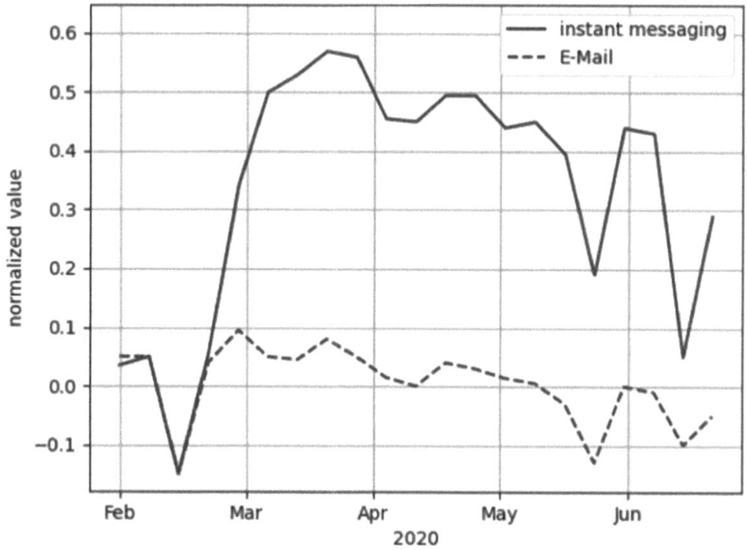

Fig. 6. Normalized numbers of E-Mails vs. instant messages in 2020 Feb–Jun at Microsoft (Yang et al. 2022)

This should be accompanied by a frequent reoccurring platform for information exchange. Especially due to the fact that ship components are interconnected and influence each other, changes that affect other sections need to be communicated with reduced delay to lower the iterations. Daily short meetings can help remotely working people inform colleagues regarding obstacles and provide insight into each other's current topics, increasing transparency. This also helps to avoid silos and ensure information is always sufficiently delivered, even with the risk of repetition.

To summarize the impact of remote working on the ship design, the influences can be decomposed into social, communication, and organizational aspects.

The lack of social aspects is mentioned as one of the main drawbacks of remote work. However, since remote working is appreciated by the vast majority of workers and will be adopted by companies, and since the advantages of remote work outweigh the disadvantages by margin, it is obvious that measures should be considered to reduce the negative effects of those disadvantages. E-Mail shall remain as the formal information exchange channel, but other channels with lower thresholds, such as instant messaging, must be not only enabled but also promoted.

Organizational workflow shall be adapted to fit better the remote working reality. For ship design, frequent meetings can increase social interactions and more transparency within the progress and avoid isolation. Furthermore, shorter portions of workloads help to keep workers motivated and increase the traceability of project progress.

Challenges for Ship Design

Ship design relies on the increase of information generated in each design iteration and the reduction of uncertainties. The reduction of design varieties throughout the loops eventually converges to the final product.

Ship systems are decomposed into subsystems and handled simultaneously and yet separately by the shipyard's design sub-departments. These departments are typically driven by the functional purpose of ship subsystems, resulting in organizational structures with, e.g., HVAC, Machinery, Ship Structure, etc. (Tsuneishi Holdings Corporation 2021; Kambase 2020; Chirillo et al. 1986; Holmes 1959). Frequent communication and information exchange across the departments are required to ensure a positive and efficient collaboration environment and, subsequently, a functional ship design.

Shipyards intended to maintain or even increase their productivity under the circumstances of remotely working employees should, therefore, focus on the two previously mentioned factors: adopt their mindset and internal processes to and implement information technology suiting the remote working models.

4 Agile Management with Scrum

Agile management is a project management approach that has gained more popularity in several fields of project engineering. Whereas traditional project management, e.g., the waterfall model, relies on planning in advance, defining deliverables, and assigning the time scope, agile management's principles consist of flexible scope, dealing with uncertainties and modifications, and delay design freezes as much as possible (Pedro Serrador and Jeffrey K. Pinto 2015).

A main strength of agile methodologies is that they better cope with deviations along any project. A deviation, in this case, is defined as a situation that occurs during a project which deviates from the original plan of this project. Since deviations are considered normal and cannot be avoided, agile assumes that the key to successful project planning lies in the utilization of tools and methodologies that can cope with and resolve these deviations instead of going with more refined initial planning to avoid them (Pedro Serrador and Jeffrey K. Pinto 2015).

Similar to the general ship design, agile is considered iterative and incremental. A major focus lies in the avoidance of design specification freeze at early design stages (Pedro Serrador and Jeffrey K. Pinto 2015). Hence, the agile approach adopted to ship design is a promising option.

Scrum
See Fig. 7.

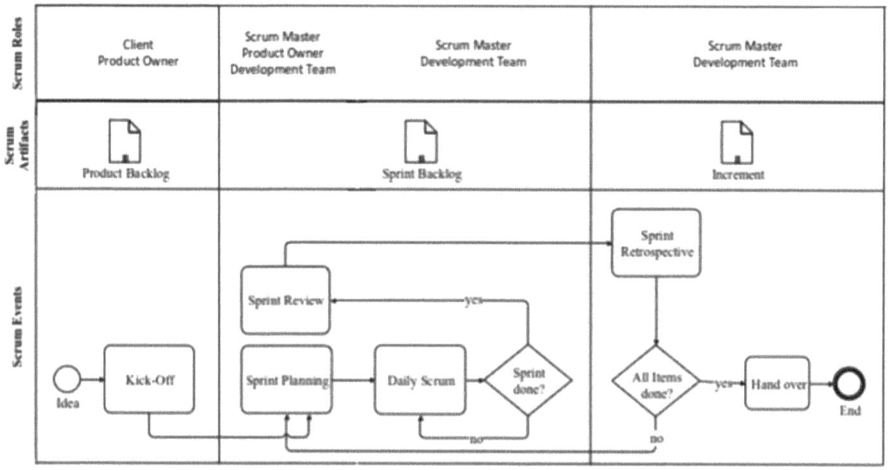

Fig. 7. Scrum Framework components (Streule et al. 2016)

Artifacts
Artifacts are defined elements with a dedicated function.

The *Product Backlog* contains the action items that need to be accomplished to form the final product. These items are ordered based on their prioritization for the product. The prioritization is determined based on various factors such as cost, resources, value, and item inter-dependencies (Charanpreet kaur and Vijay Kumar 2015).

The definition of these items is done by the product owner, who exclusively takes the perspective of the customer. Hence, the technical complexity and required effort of each item are not considered (Schwaber and Sutherland 2014; Streule et al. 2016).

Aside from a title, each item contains a value referring to its prioritization in the product backlog, a description of this item to provide a contextual frame alongside a definition of "done".

These items are further decomposed into tasks that are ideally processed by a single person. To derive tasks from items reasonably regarding the scope, scrum suggests an estimation approach to assess the complexity of the items and tasks and, if necessary, further decompose them.

The *Sprint Backlog* has selected items from the Product Backlog. These items are processed in the corresponding sprint and the amount is limited by what the product owner and the team believe can be finished within this sprint. The estimation is conducted with a complexity level for each item instead of any time-related values, even though the transfer between complexity and time consumption is given automatically by gathering experience in Scrum.

Once a task is finished and reaches its state of done, it goes into the *Increment* list, which contains all finished tasks. Since the entire scrum is timely divided into sprints, each sprint should offer an increment that provides an encapsulated functionality of the product (Schwaber and Sutherland 2014).

Roles

Scrum Roles involve all persons in the scrum process. Since each team serves a different purpose, it is crossfunctional and, meanwhile, self-organized.

The product owner takes the perspective of the customer. He is responsible for defining the product and its value according to the customer's needs/requirements. Hence, the items in the product backlog are subsequently created by the product owner. Furthermore, prioritizing and updating these items are also his responsibilities.

The development team consists of the individuals who process the product backlog items. They are specialists in the field of expertise required to design the product.

The scrum master keeps track of the work process of the entire team to ensure the scrum workflow (Schwaber and Sutherland 2014).

Events

Scrum-specific events are created to preserve a periodic frame. Consequently, any event is defined for a boxed time span. The time span is fixed for the entire project and may not be adjusted.

The Kick-off is similar to other project management frameworks and marks the start of a project. The sprint represents the core of Scrum. It is a defined time period in which a defined functionality of the product is accomplished. To ensure this structure, the sprint contains several sprint events:

- sprint planning
- development work time
- daily scrums
- sprint review
- sprint retrospective

Aside from providing time guidance, these events ensure transparency and provide the opportunity to inspect the current sprint state and to react if required (Schwaber and Sutherland 2014).

The *sprint planning* serves to plan the upcoming sprint. Here, the scope of what is delivered by the end of the sprint is set. Based on that scope, the items that will be addressed within this sprint are selected from the product backlog. The prioritization of the product backlog items is taken into account for this planning, as well as the workload, ensuring the amount of work can be done within the defined sprint time frame. The result of the sprint planning is the spring backlog, which contains all items to be completed in the sprint (Schwaber and Sutherland 2014).

The *development work time* is the actual time when the sprint backlog items are processed, sometimes just referred to as *sprint*. A sprint has a recommended maximum time frame of a single month. During the sprint, a short daily meeting, the *daily scrums*, is held by the development team and the scrum master. This meeting serves the purpose of tracking the progress of each team member and their next steps to achieve the overall sprint goal. Moreover, it provides a platform to share obstacles and difficulties among all team members and ensure transparency among all individuals (Schwaber and Sutherland 2014).

The *sprint review* is held by the end of a sprint. The accomplished items in the increment are reviewed. Based on the outcome, adaptation to the product backlog may be made based on the additional generated information during the sprint. While the sprint review is the process of reviewing the product, the *sprint retrospective* is for the team. It offers space to recap the sprint and its conduction. It summarizes the good running part and the not-good running part to detect potential improvements for upcoming sprints (Schwaber and Sutherland 2014).

Estimation

The estimation is set to evaluate the complexity level for the items or the decomposed tasks of the item from the backlogs. The common assumption in agile regarding the accuracy of estimations is that a precise prediction cannot be achieved. Therefore, only small efforts for the estimation are intended with vague results (Cohn 2005).

The evaluation is done by every member of the development team, disregarding the field of expertise. The complexity is measured on a defined scale. A frequent suggestion for the scale follows the Fibonacci sequence [1, 2, 3, 5, 8]; another one is [1, 2, 4, 8]. Both scales are non-linear to reflect the higher uncertainty in more complex items. Depending on the team, a maximum allowed complexity is defined to identify too complex items or tasks and further decompose them when required.

Within this approach, the three most common techniques for estimation are involved: expert knowledge, analogy, and disaggregation. Expert knowledge is integrated by team members who have the specific knowledge for the item. Analogy is introduced by people who cannot rely on their personal experience and therefore pull comparisons with other items/tasks and verdicts on the level of "higher" or "lower". Disaggregation suggests the decomposition of certain items/tasks when the complexity estimated before is higher than the previously defined limit (e.g., 5). Tasks are further simplified and, therefore, easier to estimate (Cohn 2005). The complexity of items/tasks is required later to define the maximum number of items processed in a sprint. In scrum, the *velocity* of a team is measured by the number of accomplished items per sprint (Cohn 2005). It is also possible to measure velocity by the numbers of complexity per sprint, which provides a more refined measurement (Moreira 2013).

Given the current situation of volatile boundaries and uncertainties in the ship design, the scrum approach designated to cope with those aspects shall be considered a promising solution. With growing experience in estimation, it will consequently become more refined and accurate. This will help the team precisely predict the duration of future projects.

5 Scrum Ship Design

The proposed agile ship design maps the conventional ship design procedure into the agile framework. In fact, it inherits the given contextual structure of the conventional ship design from the design spiral since its basic form is derived from the functional units representation of the product ship.

First, an exemplary product backlog is presented with items taken from the ship design. Afterward, a decomposition of an item from the product backlog is performed into tasks designated for a sprint.

Product Backlog

Following the definition of the product backlog, all ship functional aspects must be covered when all action points from the product backlog are accomplished. Each item's title and description should be formulated in a way that enlists all requirements from the owner's perspective. In addition, all relevant information to proceed with this item should be gathered for the development team in the description. However, the desired outputs from the ship design for the ship production are drawings (e.g., line plans, general arrangement plans, steel drawings) and machine specifications. Therefore, these outputs must be clearly described and the definition of "done" is formed accordingly.

The Product Backlog is based on the intended result from the preliminary design phase. The preliminary design phase is considered to be an early stage. Basic information is generated from the owner's mission requirements and specifications in the concept phase and taken as input. Main ship technical and economical characteristics are set (Papanikolaou 2014).

In ship design, prioritization follows the logical connection between the gathered information that is covered in one certain item and required for another item. Before the hull shape can be determined, the ship main principles such as length, beam, and draft must be known. Hence, the ship main dimensions item is placed before the hull form. Derived from this, the prioritization of the product backlog can be as follows:

1. ship main dimensions
2. hull form
3. specification of propulsion system
4. specification of auxiliary machine
5. general arrangement
6. ship structure
7. hydrostatics and ship safety

The first to be processed item is the *ship main dimensions*. With the determined values of ship length, beam, draft, side depth, displacement, and the form parameters, the *hull form* can be drawn. The actual hull shape is required to calculate the resistance of the ship during the voyage, which is mandatory to calculate the required power of the *propulsion system*. The selection of the *auxiliary machine* is based on the propulsion system. With the given hull shape and the dimensions of the machines, a *general arrangement* can be set. Since the weight distribution of each component is now given with the general arrangement, the *ship structure* can be calculated, and subsequently, the *ship safety* calculations can be performed.

The sequence given here can vary depending on the ship design procedure applied by the design departments and depending on the ship type and given mission requirements.

Item

Each item from the product backlog must come with a description providing the contextual understanding as well as requirement information to process this item. It was recommended to follow a standardized structure for the item description. For example, an input section with the relevant input parameter enables the designer to quickly gather requirement data for this item. Crucial for each item is the definition of "done". In the context of ship design, it is recommended to define "done" by referring to the given standard within the shipbuilding industry, e.g., the steel plan or common data formats.

An example of the product backlog item *Ship main dimensions* is given in the following:

> **Title**
>
> *ship main dimensions*
>
> **Description:**
>
> The ship main principles shall be determined as the initial step. These are displacement [• ^{3}], length [•], beam [•], draft [•], side depth [•], block coefficient [1].
> The main principles mark the starting point of each design. The parameters here influence almost every aspect of ship design as well as production and operation. The length and block coefficient have a significant impact on the building cost and the resistance. Beam, draft, and side death will influence the cargo volume, stability, and maneuvering behavior. For the determination, the mission requirements are given by the owner.
>
> - Shiptype: Containership
> - Cargo capacity: 2500 TEU
> - Operation Speed: 21 kn
> - Operation Distance: 21000 Sm
>
> According to the mission requirement, the ship is intended to operate through the Panama Canal. Please take into account the maximum allowed dimensions for this facility.
>
> **Definition of Done**
>
> This item is done when the ship main principles (displacement, length, beam, draft, side depth) are determined with the owner's requirements taken into account.

The selection of the title here is directly derived from the goal of this item. In ship design, the functional aspects provide a good foundation for item titles. Some aspects from the design spiral can be initially decomposed since they encompass ship subsystems with multiple components, e.g., the Propulsion can be further divided into the propeller and powering components. Each item should not represent more than one single component in the ship.

The description provides information to put this item into the context of the product. In addition, it shall point out relevant boundaries and aspects that need to be considered. Here, the provided information can be deemed as basic knowledge of ship design. In general, the product owner has to balance between required and trivial information. Therefore, knowing the background of the development team is essential to provide meaningful information for each item.

The definition of done refers to principle ship values. In most cases in ship design, the outcome of each aspect is clearly defined and, therefore, can be directly mapped. Any established file formats, drawings, and plans that are common in ship design should be used for the definition since the interdisciplinary facet of ship design heavily relies on the clear exchange of information.

Sprint Backlog

The Sprint Backlog contains all items to be processed within one sprint. Therefore, two factors are essential when deciding to fill the Sprint Backlog at the beginning of a sprint: the duration of the sprint and the number of items.

The duration of the Sprint Backlog depends on the team's internally agreed time frame for each sprint. However, it can vary during the project. Due to the various phases in ship design with different involved disciplines and expected outcomes, it can be an option to vary the time frame to a certain extent to meet the needs of every design phase. Whereas in the concept and preliminary design phase, results are coarse and, in some cases, even intended to keep the solution room flexible and wide for potential adjustments, the plans and drawings produced in the detail design phase go into production and, therefore, must be final.

The overlapping nature of design and production in shipbuilding needs to be considered when deciding on the time frame for sprints for the later design phase when production has already started. The production time of shipyards varies from yard to yard. Since the results of design departments are required by the production, e.g., the steel drawings of certain ship sections, the production time should be the driver when fixing the sprint time.

The number of items taken from the product backlog into the sprint backlog is given by the team's velocity. Since items within ship design vary throughout the design phases and departments, and decomposition can only be done to a certain extent, the velocity should be based on the complexity level. This helps to level out differences between items that will otherwise result in different workloads for sprints.

Items taken into product backlog must be estimated with a complexity level and, if necessary, further decomposed into tasks until the complexity of each item and task is within the permitted limit.

Given an example item from the product backlog:

Title

General Arrangement - Room Layout

Description:

The ship main principles were determined and an initial hull shape is given.
Create an initial room layout to fit in the desired cargo, the propulsion and auxiliary machine, and the crew.

Definition of Done

This item is done when the general arrangement in the side view is created containing the location of the required spaces.

Before assigning the rooms for the given elements to receive the room layout in this item, the following aspects must be considered:

1. the position of the collision bulkhead in the foreship section following the international regulation (SOLAS)
2. the estimated space required by the propeller and rudder in the aft ship section
3. the dimension of the potential main engine
4. the required cargo space considering the loss of space to the underwater hull shape
5. the number of required crew members, if not given
6. the required crew space area and the size of the ship deckhouse

Given an estimation of the complexity level of one for each bullet point, this item would have a combined complexity of six, which is not represented on the Fibonacci scale [1, 2, 3, 5, 8]. In this case, the next higher number, namely eight, is assumed. An eight is too high, and therefore, this item shall be further decomposed into smaller tasks. Bullet points 2 and 3 can be combined since both are related to propulsion. Bullet points 5 and 6 can be combined as well since they are connected to the law of the flag under which the ship will operate. Bullet points 1 and 4 can remain single tasks.

Given an exemplary sprint time frame of two weeks or 10 days and a velocity of 15 complexity points per sprint, another 9 complexity points are still available in this sprint.

During the sprint, daily scrum helps to detect if the initial estimation is still valid or if further actions are required, e.g., when additional items can be added to the ongoing sprint backlog, or items need to be returned into the product backlog for future sprints because of exceeding scope. By this, any possible deviations caused by project uncertainties are detected right on time and countermeasures can be taken.

In addition, the daily scrum can serve as a daily frame for the development team to interact with each other. Though the daily scrum itself shall be reduced to its essential purpose of providing a brief update, it initializes social interactions and lowers thresholds for further communication outside of the daily scrum frame.

6 Conclusion

Recent events have led to a set of changes in ship design. International regulations to reduce GHG-emissions force ships to adopt new technologies and components. Combined with the remote working trend since the pandemic, ship design is facing a new situation with many uncertainties. Meanwhile, the foundation of the nowadays still widely applied methodology of designing a ship mostly relies on empirical data and experiences in the crucial initial phase.

To govern the new situation for the ship design, an agile management approach utilizing scrum is proposed. Agile management follows a more flexible strategy to better deal with uncertainties and deviations during the project. Scrum is dedicated to product development and is especially suitable for complex products involving interdisciplinary specialized teams. The project organization follows a strict time scheme with reoccurring events involving the participating development teams to provide transparency. Any deviations are quickly identified and possible actions can be taken on time.

The scrum approach takes the product to be developed and splits it into chunks. Each chunk represents an aspect of the final product. The chunks are prioritized and listed in the product backlog.

For the proposed scrum ship design, it was shown that the technical principles of ship design can remain since the traditional ship design spiral approach follows the functional principle of a ship. Requirements of the ship are represented by its functionalities instead of actual components. By this, an openness to technology is ensured, as long as the result ensures the functionality of the ship. Potential new technologies required by the ship environmental regulations are coped with in this way. Thus, the product backlog can be created based on the design spiral. The prioritization of entries in the product backlog follows the dependencies depicted in the ship spiral. Any item marks a blocker for the following items when not conducted in the right sequence.

Once the product backlog is created for a ship, the procedure of the scrum ship design closely follows the general scrum approach. In Scrum, the product development procedure is organized in sprints. Sprints follow a time grid, mostly between two and four weeks. A sprint backlog is created for each sprint, where the items to be processed in this sprint are taken from the product backlog. The number of backlog items for each sprint is set by the estimation of all team members.

A daily short meeting ensures transparency and traceability of the progress of the sprint and provides social interactions between team members, especially when working remotely.

The design spiral remains valid. It serves as the foundation for the proposed scrum ship design. The organizational aspects during the ship design are addressed with the scrum procedure. It helps to identify deviations during the design process, and its daily structure provides a platform for workers from a social perspective. The estimation of the sprint scope is a source of uncertainty itself, but due to the nature of scrum, any consequences can be identified during the process and coped with actions. With time going, estimations will be more accurate and tailored to the team's abilities. This will further improve the time prediction of future projects for the shipyard.

References

Angle, A., Brackley, A.: Trends - What's Next for Sustainable Business? 2022 Trends Report. The SustainAbility Institute by ERM (2022)

Burrows, M., Hadzic, D., Stoltz, K.: Current trends in remote working. Survey. KPMG International (2021)

Kaur, C., Kumar, V.: Prioritization in scrum: impact of maturity of product backlog item. Int. J. Adv. Trends Comput. Appl. **2**, 56–62 (2015)

Chirillo, R.D., Inuki, I., Kobayashi, I.: Shipyard organization and management development, pp. 47–79 (1986)

Cohn, M.: Agile Estimating and Planning. Prentice Hall, Upper Saddle River (2005). http://www.amazon.com/Agile-Estimating-Planning-Mike-Cohn/dp/0131479415

Deloitte: THome-Office Umfrage 2021. Deloitte (2021)

DNV: IMO Update: Marine Environment Protection Committee – MEPC 78. In Technical Regulatory News No. 16/2022 (2022)

Erikstad, S.O., Lagemann, B.: Design Methodology state-of-the-art report. Report. Norwegian University for Science and Technology (2022)

Errath, M., Holzinger, A., Slany, W.: Agile software Entwicklung. OCG J. **29**, 4–6 (2004)

Evans, H.: Basic design concepts. Naval Eng. J. (1959)

Germanischer Lloyd: Guidelines for Determination of the Energy Efficiency Design Index. Ship Classification Rules and Guidelines. Germanischer Lloyd (2013)

Hammer, L., Eide, M.: Maritime Forecast to 2050. Report. DNV (2021)

Holmes, D.T.: A proposed Naval Shipyard organization. Theses and Dissertations. George Washington University (1959)

International Maritime Organisation: Amendments to the annex of the Protocol of 1997 to amend the International Convention for the prevention of the Pollution from ships, 1973, As modified by the protocol of 1978 relating thereto. International Maritime Organisation (2011)

International Maritime Organisation: 2018 Guidelines on the Method of Calculation of the Attained Energy Efficiency Design Index for new Ships. Standard. International Maritime Organisation (2018)

International Maritime Organisation: Draft guidelines associated with draft amendments to MARPOL Annex VI to incorporate the goal-based energy efficiency improvement measure utilizing Energy Efficiency Existing Ship Index (EEXI). International Maritime Organisation (2020)

International Maritime Organisation: Introduction to IMO (2023). https://www.imo.org/en/About/Pages/Default.aspx. Accessed 22 Dec 2023

Kambase, N.: Ship maintenance and repair services in the Tema Drydock and shipbuilding Yard in Ghana: some challenges, opportunities and prospects. World J. Eng. Technol. **08**, 296–316 (2020). https://doi.org/10.4236/wjet.2020.83024

Khan, L., Macklin, J.J.R., Peck, B.C.D., Morton, O., Souppez, J.-B.R.G.: A review of wind-assisted ship propulsion for sustainable commercial shipping: latest developments and future stakes. In: Wind Propulsion Conference, London, UK (2021)

Lappalainen, O., Spaniol, M.J., Karvonen, T., Ehlers, V., Karlsson, J., Nåfors, D., et al.: Road to Shipyard 4.0: the state of play, a brief history of maritime developments, and a future roadmap. Focusing on the Baltic Sea and Shipyards. Report. ECOPRODIGI (2020)

Liu, S., Ang, J.H.: Challenges in meeting upcoming EEXI. In: The Society of Naval Architects and Marine Engineers-SNAME (2021)

MAERSK Press: Maersk secures green methanol for maiden voyage of the world's first methanol-enabled container vessel (2023). https://www.maersk.com/news/articles/2023/06/13/maersk-secures-green-methanol. Accessed 22 Dec 2023

McTaggart, V., McLaughlin, C.: Title: Remote working the new reality (2020)

Meyer Werft Presse: MEYER WERFT ist Vorreiter beim Infektionsschutz (2021). https://www.meyerwerft.de/de/presse/presse_detail/meyer_werft_ist_vorreiter_beim_infektionsschutz.jsp. Accessed 8 Jan 2024

Moreira, M.: Being agile: your roadmap to successful adoption of agile (2013)

Nowacki, H.: A farewell to the design spiral. In: Ship Design, Ship Hydrodynamics & Maritime Safety, Athens (2016)

Olson, M.H.: Remote office work: changing work patterns in space and time, vol. 26, pp. 183–187 (1983)

Ovrum, E., Longva, T.: Maritime forecast to 2050. Report. DNV (2022)

Papanikolaou, A.: Ship Design - Methodologies of Preliminary Design: Springer Dordrecht (2014). https://doi.org/10.1007/978-94-017-8751-2

Pearson, D.: The use of Flettner rotors in efficient ship design. In: RINA, Royal Institution of Naval Architects - Influence of EEDI on Ship Design 2014, pp. 162–169 (2014)

Serrador, P., Pinto, J.K.: Does agile work? — a quantitative analysis of agile project success. Int. J. Proj. Manag. **33**(5), 1040–1051 (2015)

Prinz, W.: Flexibles Arbeiten im Homeoffice - Analyse einer Langzeitumfrage, pp. 146–152 (2021)

Ramboll: Kraftstiffanalyse in der Schifffahrt nach Segmenten. Report. Deutsches Maritimes Zentrum e. V (2022)

Scandlines: Ferry Operator & Passengers are Moved by the Wind (2020)

Schwaber, K., Sutherland, J.: Der Scrum Guide (2014)
Serpa, L.B., Lewin, P., Molander, T.: 2022 Sustainability Report. Report. MAERSK (2023)
Streule, T., Miserini, N., Bartlomé, O., Klippel, M., García de Soto, B.: Implementation of scrum in the construction industry. Procedia Eng. **164**, 269–276 (2016). https://doi.org/10.1016/j.proeng.2016.11.619
Sutherland, J.: The scrum papers: nut, bolts, and origins of an agile framework (2012)
Tokuslu, A.: Analyzing the energy efficiency design index (EEDI) performance of a container ship. Int. J. Environ. Geoinform. **7**, 114–119 (2020). https://doi.org/10.30897/ijegeo.703255
Tsuneishi Holdings Corporation: Tsuneishi Corporate Section (2021). https://www.tsuneishi.co.jp/english/corporate/section/
Voss, L.: Hydrogen as Ship-Fuel. Report. ELMAR (2020)
Vossen, C., Kleppe, R., Hjørungnes, S.: Ship design and system integration (2013)
VSM AG: Schiffbau Industrie II (2021). https://www.vsm.de/de/service/publikation-und-medien/schiffbau-industrie
Wanf, X., Plinder, H., Negenborn, R.R.: Types of ship components and technical characteristics. TU Delft (2020)
Yang, L., Holtz, D., Jaffe, S., Suri, S., Sinha, S., Weston, J., et al.: The effects of remote work on collaboration among information workers. Nat. Hum. Behav. **6**(1), 43–54 (2022)
Ziółkowski, M., Dyl, T.: Possible applications of additive manufacturing technologies in shipbuilding: a review. Multidiscip. Digit. Publ. Inst. MDPI J. (2020)

Utilisation of Findings from a Series of Studies to Explain the Mechanisms of Action of Mechatronic Systems

Kristin Paetzold-Byhain(✉)

Dresden University of Technology, Dresden, Germany
kristin.paetzold@tu-dresden.de

Abstract. Agile methods are attracting more and more attention for the development of mechatronic systems. They promise development organisations in companies flexibility and responsiveness, enabling them to react adequately under dynamic and volatile conditions. Originally, agile methods were considered to support software development. However, they appear to be quite suitable for supporting the development of mechatronic systems. The main difference between software development and the development of mechatronic systems was initially seen as being that mechatronic systems have to materialise. Functionalities and properties of mechatronic systems can only be realistically used after they have been manufactured. Physicality is therefore considered to be the most restrictive aspect in the use of agile methods for mechatronics development. More detailed analyses show that the high level of functional integration and the complexity of the development in particular also present a challenge. Division of labour for large complex systems is necessary to be efficient in development. Development tasks are not independent of each other. This requires a rethinking of agile working mechanisms to provide the best possible support for the development of mechatronic systems.

Another interesting phenomenon in the use of agile methods is that they were initially taken up from practice. Many companies have experimented with agile methods, which is reflected not least in the fact that large application-focused conferences on agile work quickly became established. However, this is also accompanied by a so-called guru problem. Individual persons or companies demonstrate the successful application of the methods on the basis of best practices. If these are used without reflection, they do not necessarily lead to success, especially if the application contexts differ greatly. Although it is clear from the best practices what works in the application of the method, this is usually only partially explainable because descriptive cause-and-effect mechanisms are lacking. As a consequence, the introduction of the method fails. This means that effective and promising approaches are easily burned.

Consequently, the paper was based on the following research questions: How can cause-and-effect mechanisms and means-end relationships of agile development be identified, and how must agile methods be adapted to make them effective for the specific characteristics of mechatronics development and for specific application contexts? The article summarises the findings from a series of studies, from which implications for the use of elements of agile working, in this case prototyping in particular, are derived and explained.

K. Gericke et al. (Eds.): DSC 2022/2023, PTT, pp. 113–129, 2026.
https://doi.org/10.1007/978-3-032-16278-6_8

1 Introduction

Agile methods are gaining more and more attention for the development of mechatronic systems (Komus et al. 2020), as they promise the development organizations in companies flexibility and responsiveness in order to be able to react adequately under dynamic and volatile conditions (Dingsøyr et al. 2012). Agile methods were originally considered to support software development. However, they certainly appear to be suitable for supporting the development of mechatronic systems (Conforto et al. 2014). The main difference between software development and the development of mechatronic systems was initially seen in the fact that mechatronic systems have to materialize. The functionalities and properties of mechatronic systems can only be realistically utilized once they have been manufactured. Physicality is therefore considered to be the most limiting aspect in the use of agile methods for mechatronic development (e.g. Atzberger et al. 2019).

Another interesting phenomenon in the use of agile methods is that they were initially taken up by practitioners. Many companies have experimented with agile methods, which is reflected not least in the fact that large application-focused conferences on agile working quickly became established (e.g. Agile Pep-Minds). However, this is also accompanied by a so-called guru problem. Individuals or companies demonstrate the successful application of methods based on best practices. If these are used without reflection, they do not necessarily lead to success, especially if the application contexts differ greatly. Although it is clear from the best practices what works in the application of the methods, this can usually only be explained to a limited extent because there are no descriptive cause-and-effect mechanisms. As a result, the introduction of the method fails. This means that effective and promising approaches are easily burnt.

Consequently, our research work was based on the following research questions:

How can cause-and-effect mechanisms as well as means-and-purpose relationships be used in agile development and how must agile methods be adapted in order to make them effective for the specific characteristics of mechatronics development and for specific application contexts?

2 Understanding of Agile Methods

Agile development methods were originally considered for software development. The basis for this was the agile manifesto, the basic statements of which are shown in Fig. 1.

With these arguments, the authors are primarily addressing an imbalance in development. The importance of the aspects on the right-hand side (black boxes) is well recognized, but the aspects on the left-hand side (grey boxes) are perceived as adding more value. Sutherland (2014) explain that excessive planning in situations where the ability to forecast is limited binds massive capacities of developers for technical management tasks, which are lost for the actual creative-creative development. Non-value-adding activities in development must be minimized in favour of effective and efficient development work. However, management tasks remain necessary, especially in mechatronics development, in order to bring together the results of the division of labour, but also to serve the information needs and interfaces required by the high degree of functional integration (Kossiakoff et al. 2020).

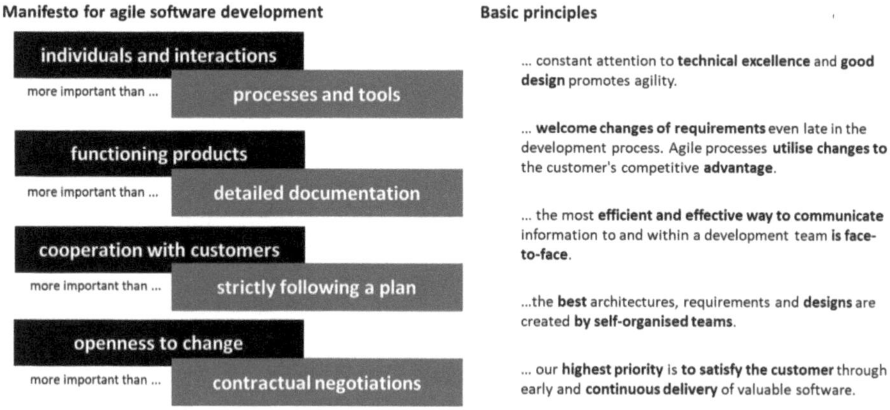

Fig. 1. Value system of agile development (Beck et al. 2001)

To implement the philosophy of agile development, 12 principles have been defined (see Fig. 1, right-hand side) (Beck et al. 2001), which should guide concrete action. These basic principles emphasize the necessity and importance of technical excellence, thereby focusing on the actual problem solution. By postulating the acceptance of change, a new way of dealing with moving targets is emphasized. Time-consuming and costly risk assessments are eliminated in favour of change management. In addition, the role of developers in information flows is strengthened, which at the same time forms the basis for self-organised action and transparent decision-making processes (e.g. Dyba et al. 2008; Gustavsson 2016). Last but not least, the intensive involvement of customers in development not only emphasizes their expertise in the usage situation and takes this into account in decisions. Continuous validation with the customer also gives the developers the opportunity to discuss new technologies with the customer to find solutions and to evaluate their potential holistically.

Based on the agile manifesto and the basic principles, a number of process models and methods for agile working have ultimately emerged. Examples include Scrum (Sutherland et al. 2020), Extreme Programming (Beck 2000), Kanban (Anderson 2010) and Design Thinking (Gonen 2020). These prioritize specific aspects of development, but essentially follow the same values. In summary, agile methods are characterized by the following features (Paetzold 2022):

- Incremental development allows a reduction in complexity through task decomposition and frequent validation of requirements
- Iterative approach leads to higher verification frequency for intermediate results.
- Teamwork simplifies coordination and supports communication and guarantees the necessary knowledge for finding solutions
- User/customer integration reduces misunderstandings and misinterpretations of requirements.

One noticeable difference to the classic development approach lies in the target planning and interpretation. In traditional development projects, process planning is carried out before the start of development with the aim of minimizing deviations from the defined objectives (requirements) and the planned process. Plans should be followed as faithfully as possible in order to achieve initial goals as efficiently as possible. Iteration points (such as milestones) are then mainly used for verification (Schrof et al. 2020) (Δu). This approach is not least due to quality thinking, according to which it is postulated that the adherence to and consistent implementation of process steps brings with it a certain completeness in the consideration and thus helps to avoid errors in the product (Brugger-Gebhardt et al. 2016). The focus is therefore always on a pronounced risk assessment. Agile methods also follow structured processes that reflect the development logic. However, the focus is also on minimizing the difference between the planned procedure and the reality, which is characterized by uncertainties (Δz). Parallel to the development process, the objectives are now validated in addition to the verification of the implementation. In this way, uncertainties are accepted and instead perceived as opportunities. The different ways of thinking are illustrated in Fig. 2.

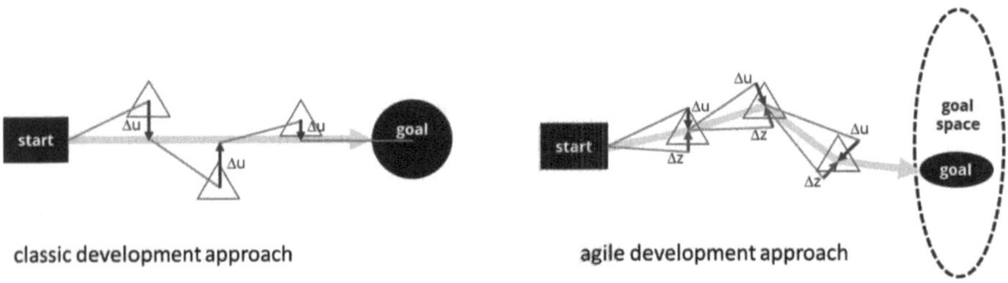

Fig. 2. Comparison between the classic and agile approach (according to Schrof et al. 2020) (Δu describes deviation from the planned process; Δz describes deviation from planned targets)

One of the best-known and most widely used agile approaches is Scrum (e.g. Sutherland et al. 2020). Scrum is to be understood as a process model that follows the Plan-Do-Control-Act management cycle (e.g. Stelzmann 2011, according to Boyd 1996), which is also used in project management. Scrum is a constituent and operationalizing methodology whose emergence and development is driven by application, is based on empiricism and describes empirical knowledge (Sutherland 2014).

The process model behind Scrum is summarized in Fig. 3. The cycle begins with sprint planning, which is based on user stories or utilizes the product backlog (Plan). During the actual sprint, the defined development task is worked on (Do). Here, the development teams use their familiar methods and tools for synthesizing and analyzing. This follows the problem-solving cycle according to (Ehrlenspiel et al. 2013). On the one hand, the daily meetings serve as a control element, as not only the current work status is stated here, but also possible problems and challenges are discussed. Daily meetings are therefore not only used for reporting, i.e. the pure exchange of information. Rather, they form the basis for initiating changes and reacting to uncertainties. This corresponds to a kind of internal control. At the end of a sprint, the implemented functionalities (increment) are presented to the customer, product owner and other stakeholders and discussed with them in a sprint review meeting. This creates further opportunities to recognize problems and obstacles and initiate changes. The feedback is then incorporated as new or adapted requirements. Conversely, it enables the development team to remain agile, minimize waste, invest resources in an outdated plan and ultimately provide the customer with a product tailored to their needs. This is an external control loop. Sprint retrospectives, which take place after a sprint review but before the next sprint planning phase, give the team the opportunity to reflect on their work and serve as a learning process within the team. The assignment to the management approach (PDCA) is colour-coded in Fig. 3.

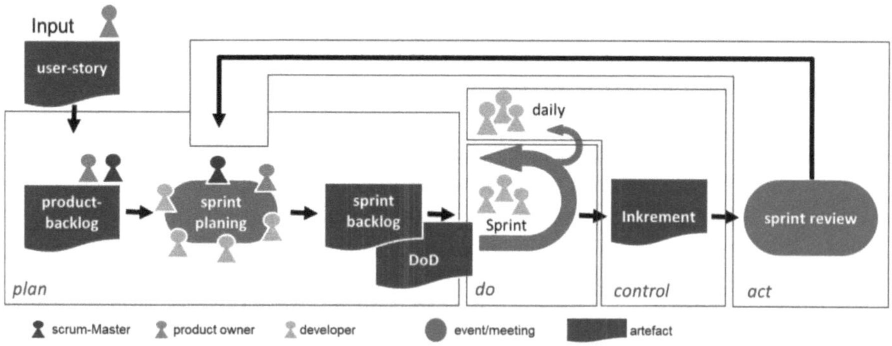

Fig. 3. Scrum as a process model for agile working (according to Sutherland et al. 2020)

The mindset behind agile methods, as illustrated by the Scrum approach, has certainly proven to be a promising way to act in a goal-orientated manner in dynamic and volatile development environments. There are two challenges for widespread application:

- Agile methods are initially intended for the development of software products. Due to their virtual nature and the way in which they are structured (e.g. Boehm 1988), certain elements such as the conversion of increments into continuously realizable products are easier to organize than is possible with mechatronic products, as their full functionality is usually tied to their production (Ovesen 2012).
- Agile methods are designed for so-called sweet spot conditions (Kruchten 2013). These describe that development takes place in small teams of 10–15 people at a common location, the customer is available and the development work has a high proportion of new developments, is based on short development cycles of a few months and, above all, there is a common understanding of the processes (Kruchten 2013, p. 355). The project conditions differ significantly from development tasks in mechatronics development, where a large number of sometimes distributed development teams contribute to task fulfilment and the customer is often not available for feedback. As a result, problems are usually solved by negotiation. Long development times are the rule for large, complex systems. In addition, the development environment (development organization, but especially IT structures in development) determines the task processing (Atzberger 2021).

One of the main challenges in the use of agile methods for the development of mechatronic systems is seen as their physicality, not least because of the two aspects mentioned (e.g. Rößler et al. 2022). In (Ovesen 2012), Ovesen analyses the challenges and difficulties of agile development in the context of integrated product development. From the analysis of development work in large development organizations, he confirms the bitter spot conditions mentioned by Kruchten (Kruchten 2013) and identifies and clusters the challenges in agile development. In addition to Paradigm Perplexity, Designers Dissent, Team Distribution Dilemma and Necessity of education and maturity, Oveson also identifies the Constraints of physicality. However, analyzing the individual challenges in the application context quickly makes it clear that they are not independent of each other. For this reason, the challenges mentioned were used as the basis for the analyses of the mechanisms of action of agile methods and for theory development.

3 Methodological Approach

In addition to the theory-building development methodological investigations, studies on the application of agile methods in industry should not only contribute to data collection (Weiss et al. 2023). Since the use of agile methods was initially strongly promoted by industry, it seems obvious to take up the experience gained here. The aim is to capture best practices, but also to identify challenges and solutions and take them into account in the development of theory. A series of studies was developed for this purpose, which is presented below. The study "Agile development of physical products" was launched in 2017 and is now in its 6th edition (Weiss et al. 2023). The series of studies can be used to show how experience from industrial application helps to record, describe and validate mechanisms of action.

3.1 Basic Structure of the Study Series

The study initially started with the aim of analyzing expectations and challenges in dealing with agile methods in order to understand the capabilities and potential of agile methods. In the course of the study, it was then increasingly used to test mechanisms of action developed from theory and to derive implications for the mechanisms of action from practice. The hypotheses were based on the challenges identified by Oveson (2012) for integrated product development (see Sect. 2), whereby the physicality of mechatronic systems was initially seen as the strongest driving cause of challenges.

The study comprises general questions on the use, understanding and application of agile methods, in most cases with answer options according to a five-point Likert scale. The sample size varies between 250 and 150 participants over the years, primarily from Germany and the so called DACH region (Austria, Switzerland, Germany). This focussing of the sample seems important in order to be able to exclude country-specific cultural differences in working methods. In recent years, it has always been very successful in representing a broad mix of industries, as can be seen in Fig. 4. We also asked about the size of the companies in which the participants are employed.

Demographic data, which is collected in the same way every year, records the position of the participants in the company, the distribution of participants by field of activity and the distribution of participants according to their experience in dealing with agile methods (see Fig. 5). Participants who specified pure software development as their field of activity were not included in the analysis, as the focus is on experience with agile methods for physical products. On the one hand, it is clear from the diagrams that the

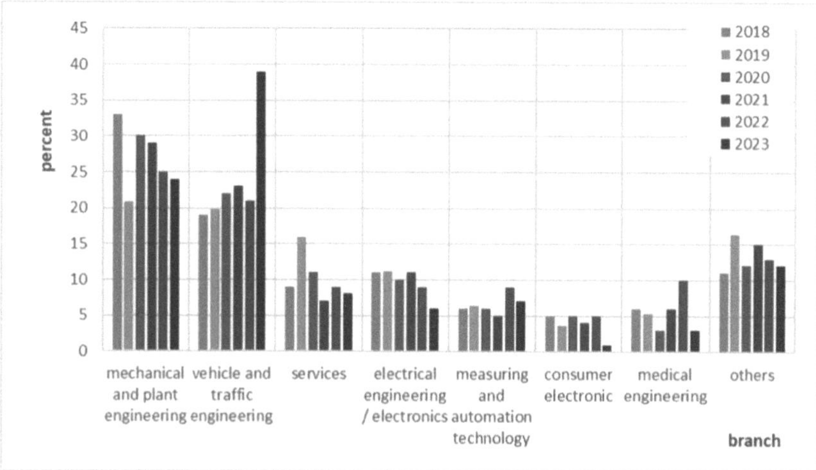

Fig. 4. Cumulative presentation of sector distribution over the years

high proportion of developers gives us a deep insight into the product development of the participating companies (Fig. 5, top left). It was essentially possible to gain insights from two perspectives: from the perspective of the developers themselves and from the perspective of management or project responsibility (Fig. 5, top right). The diagram in Fig. 5 below shows a slight trend towards increasing experience in the use of agile methods for the development of mechatronic systems.

3.2 Hypothesising with a Focus on the Physicality of Mechatronic Products

The first two studies initially scrutinised the benefits and challenges of agile methods for mechatronics development. The starting point for the surveys was a list of items that were associated with the benefits and challenges of agile working. The items on the list were based on extensive literature studies, real implementation projects and assumptions made by our research group. The randomised order of the items from more "hard" economic aspects and "soft" team and collaboration effects was intended to help ensure objectivity in the responses. The aim of the first surveys was to identify actual potential and specific challenges for mechatronics development in order to initially derive means-purpose relationships from this.

These questions on the benefits of agile methods were repeatedly addressed in subsequent studies in order to derive trends in the dissemination and adaptation of agile methods. In addition, questions were added for various sub-aspects, such as the handling and understanding of prototypes, scaling, the introduction of agile methods, the evaluation of success and, last but not least, the associated transformation process. Hypothesizing was based on the challenges according to Ovesen (2012). Initially, the focus was on physicality. However, it quickly became apparent that Ovesen's criteria are not independent of each other. The diverse interactions were taken into account when formulating the questions and defining the items.

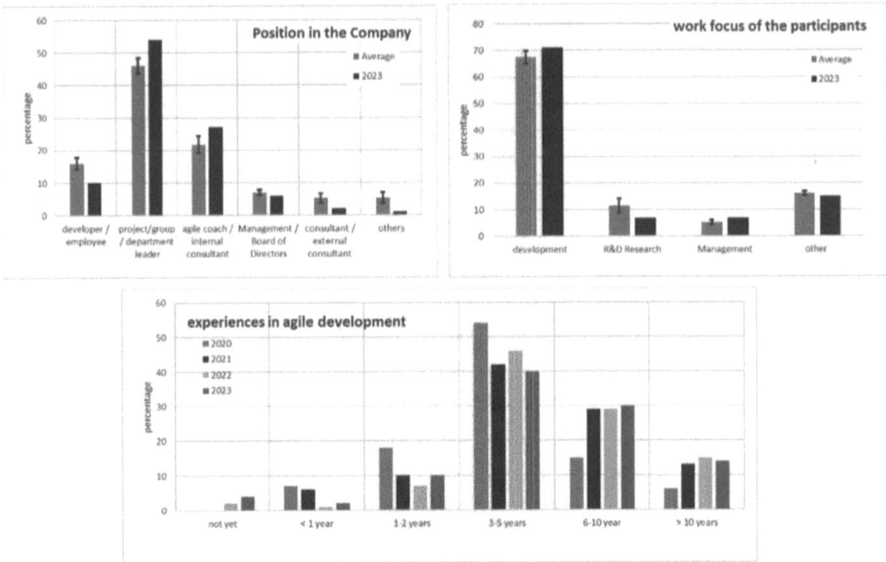

Fig. 5. Demographic information about the participants

4 Selected Results from the Study Series

Selected results on the benefits and challenges of agile working and prototyping are presented below. The main aim of the studies, particularly in the early years, was to determine the potential of agile methods for mechatronics development. The physicality of the systems to be developed was emphasised as an essential challenge for their use, both in industry and from a scientific perspective. It was also initially observed that physicality was very strongly linked to the requirement for prototypes. The intensive use of prototypes, especially as a communication tool, emphasises a task that deviates from the usual understanding of prototypes (e.g. Hallgrimsson 2012) and can therefore easily lead to misunderstandings. Since the benefits of continuous communication with the user via the prototypes were also emphasised, the initial research focused on understanding the effects of physicality when using agile methods in the development process. It was important for us to find out how to identify causal chains linked to this for agile development and to understand the mechanisms of action when using agile methods. I hope it works.

4.1 Challenges of Agile Methods for Mechatronics Development

As already described, the study initially questioned expectations and challenges regarding the use of agile methods for the development of mechatronic systems. The diagram in Fig. 6 compares the mean values of the actual challenges with the expected ones. In addition, the difference analysis allows a statement to be made about the aspects in which agile methods were over- or underestimated. While the statements on expectations contain the responses of all participants, the diagram for the actual challenges does not include the responses of participants who have no experience or are just starting to use the methods.

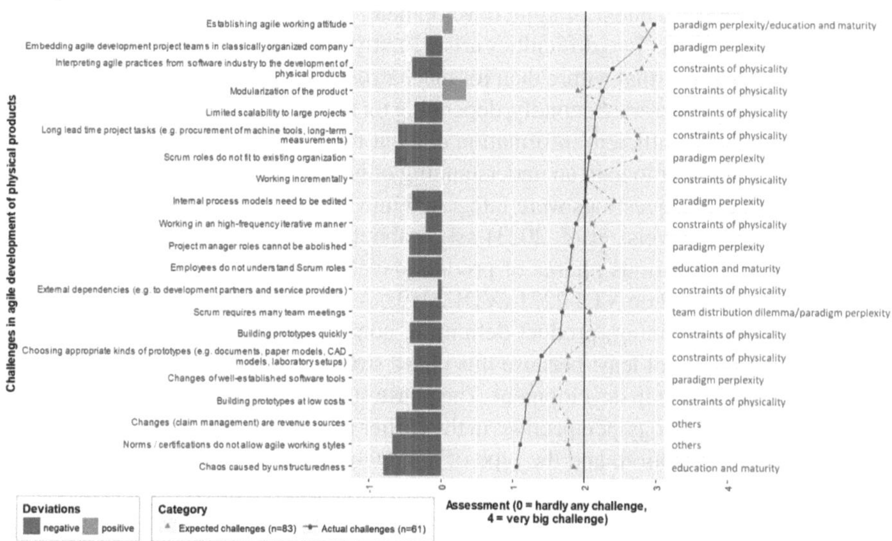

Fig. 6. Actual and expected challenges in the use of agile methods for mechatronics development (Weiss et al. 2023/2018)

It is clear from the diagram that none of the challenges mentioned are seen as unmanageable. On the 5-point Likert scale, the maximum values are around 3). Accordingly, the establishment of a suitable mindset, the embedding of agile teams in classically organised development organisations and the interpretation of practices from software development are seen as real challenges. Although challenges resulting from physicality are recognised, they are obviously not seen as a major obstacle to the use of the method. It is interesting to note that most of the challenges tested were overestimated (see difference analysis).

Many of the challenges result from organisational structures that have grown in companies and been established over the years. Behind this, of course, are practised working methods and established task divisions. Teamwork is also closely linked to this. This is well established in traditional development organisations, but is generally not based on the idea of collaboration, but rather on cooperation, which can also be used to justify centralised decision-making processes. Such aspects must not only be taken into account when adapting and interpreting agile methods, but also influence questions of knowledge and information distribution, which in turn is influenced by physicality.

4.2 Results for the Evaluation of Prototyping

He series of studies will also analyse details of agile development in order to be able to describe mechanisms of action in detail. In discussions with industry partners, the topic of prototyping is repeatedly the subject of controversial debate. Looking at agile process models, such as Scrum (see Sect. 2), an increment is expected at the end of a development cycle, which fulfils the goals defined for the sprint (Definition of Done) and completes the resulting product. In this context, an increment is usually interpreted as a prototype. In the context of agile development, prototypes are used for validation but also for exploration. Two difficulties arise. Firstly, prototyping is easier for software products due to their virtual nature than for mechatronic systems, for which materialisation appears to be necessary. Secondly, depending on the division of tasks, increments do not necessarily have sufficient information content to meaningfully depict and discuss functionalities. In order to gain an understanding of how prototyping is to be interpreted in agile development, questions were integrated into the study to understand it.

The 2022 study (Weiss et al. 2023) asked about the use of prototypes (frequency and purpose) in the various phases of product development. The development phases were categorised based on VDI 2221 (VDI 2019). A distinction was made between the phases of product planning, product definition, concept development, series development and validation, not least because this phase structure often forms the basis of the organisational structure in development. The phase structure can also be justified from a development methodology perspective: in the earlier phases, it is more important to deal with uncertain information, and the type of uncertainties also changes over the course of the development process (Walker et al. 2003). This leads to different evaluations of prototypes.

Figure 7 above shows the frequency with which prototypes are used. The bottom shows how helpful the use of prototypes is rated in the development phases. It is clear that prototypes are obviously used as an effective tool for concept and series development as well as for validation, but not in the product planning and definition phases. However, more than half of the participants rated the use of prototypes as helpful, especially for product planning and definition.

The understanding of prototyping in agile development is based on two assumptions: On the one hand, prototypes serve as a means of communication to validate development results with the customer. On the other hand, prototypes support exploration (Punkka 2012). Testing on prototypes provides the opportunity to explore and understand interdependencies. The understanding of prototypes associated with agile development differs somewhat from the classic understanding (e.g. Budde et al. 1990). In order to better utilise this understanding, we wanted to know from the participants which tasks are usually supported by prototypes. The results are shown in Fig. 8.

The grouped bars show that the participants generally rate the use of prototypes as recommendable, both for validation and communication tasks as well as for verification. It is interesting to note that the approval rate for the use of prototypes is low for the exploration of ideas or the testing of (market) hypotheses. Both are fields of application in the early phases of development.

In order to understand more precisely for which activities the prototypes are used in development, the participants were given the statements listed in Fig. 9 for evaluation. This also shows that the importance of prototypes in an agile context is fundamentally understood. This is manifested in the high percentage of agreement. Difficulties are particularly evident in the categorisation of product artefacts in agile development. It

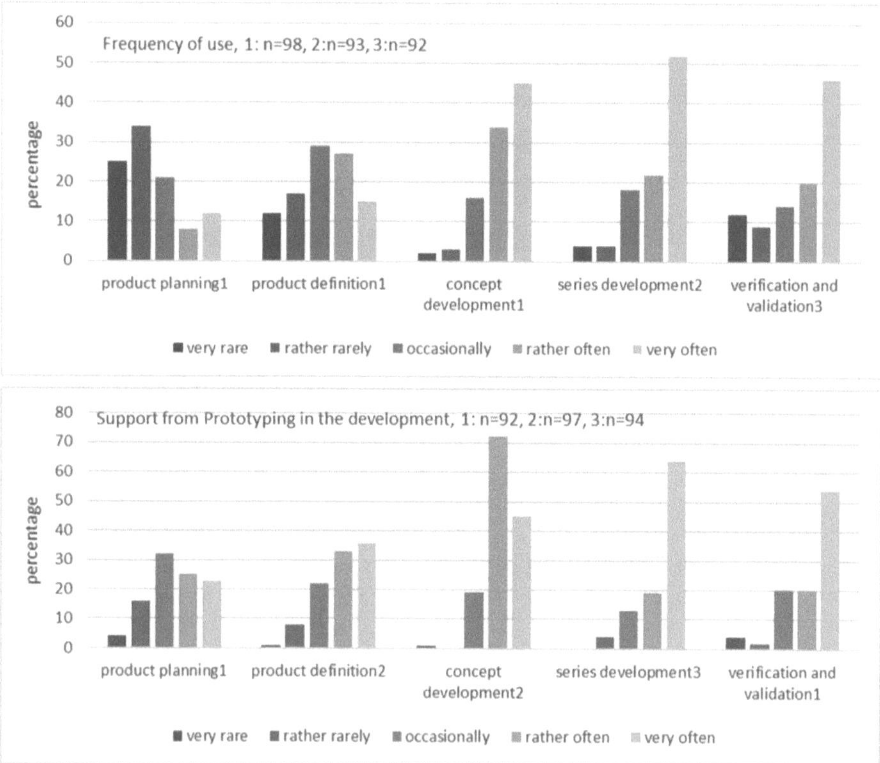

Fig. 7. Assessment of the frequency (top) of and support by prototypes in the development phases (bottom) (Weiss et al. 2023/2019); 1: n = 92, 2:n = 97, 3:n = 94

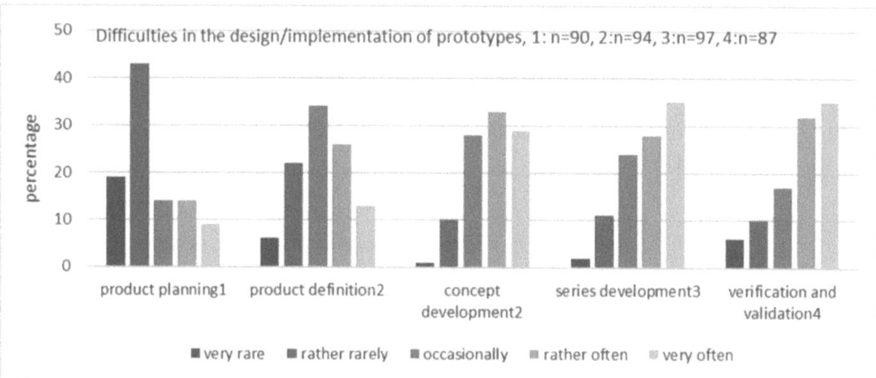

Fig. 8. Difficulties in the design/implementation of prototypes in different development phases (Weiss et al. 2023/2019)

is evidently difficult to clearly differentiate between iteration, increment and prototype, which can be deduced from the rather low agreement with the latter three items (Fig. 10).

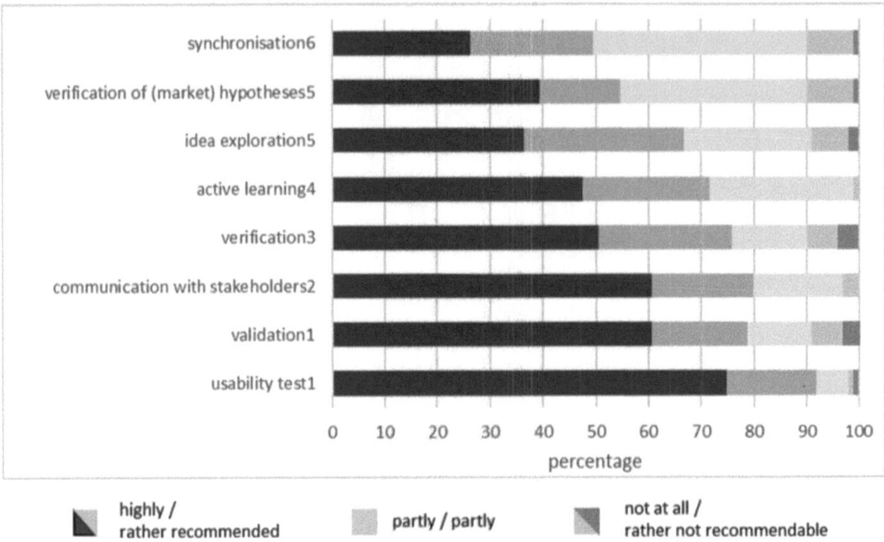

Fig. 9. Recommendations for the use of prototypes for defined applications (Weiss et al. 2023/2019); 1: n = 100, 2:n = 101, 3:n = 96, 4:n = 99, 5:n = 98, 6:n = 94

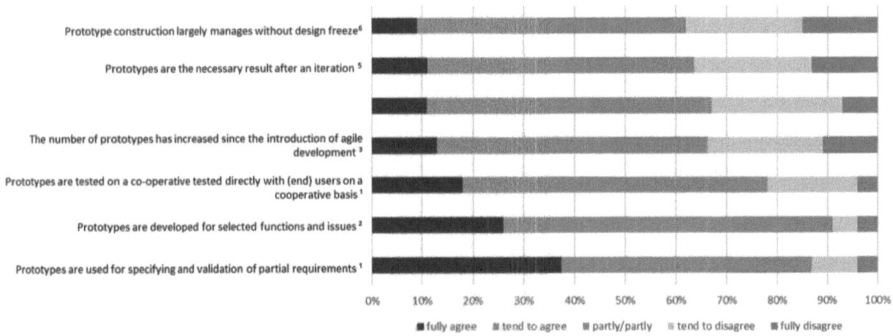

Fig. 10. Agreement with the above statements (Weiss et al. 2023/2019); 1: n = 100, 2:n = 99, 3:n = 92, 4:n = 97, 5:n = 99, 6:n = 96

When using agile methods for mechatronics development, it is repeatedly argued that physicality in particular hinders rapid prototyping. The 2020 study (Weiss et al. 2023/2020) therefore asked about the tools used to realise prototyping. Figure 11 shows the results.

Additive technologies and document-based prototyping are obviously standard in the development of mechatronic systems. The use of vertical prototypes, i.e. the visualization of a reduced product scope, is also common. It is also interesting to note that virtual product artefacts are seen as adequate for prototypes. Originally intended for analysis purposes, digital models are also used for the prototypical visualization of product artefacts. As far as digitalization is concerned, the use of relatively new technologies such as AR/VR or the use of digital twins is not yet the norm - they are only used in just

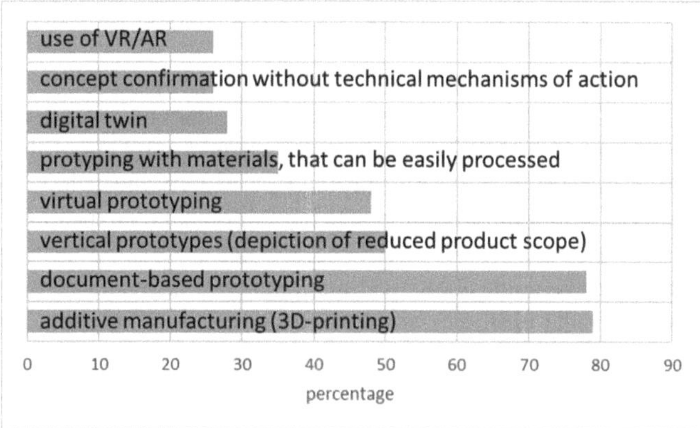

Fig. 11. Tools used for rapid prototyping (Weiss et al. 2023/2020) (multiple answers possible, n = 95)

over a quarter of companies. Visual samples, surface models and "fake prototypes", i.e. concept confirmations without technical mechanisms of action, are obviously not part of the standard repertoire for most companies either (Weiss et al. 2023/2020).

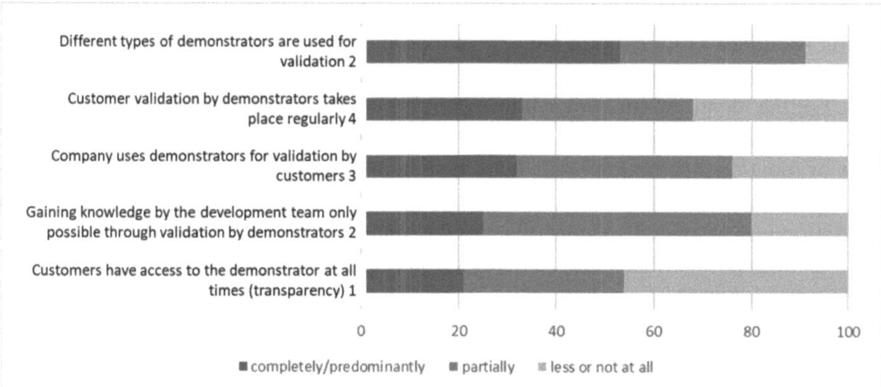

Fig. 12. Approval ratings regarding experiences with customer validation during the development process (Weiss et al. 2023/2023); 1:n = 57, 2:n = 80, 3:n = 78, 4:n = 79

Results from the 2023 study (Weiss et al. 2023/2023) confirm that companies find workarounds for prototyping for validation with customers. In addition to physical demonstrators, virtual demonstrators were also proposed as response options for demonstrators used. In addition, demonstrators from different development phases were also included in the evaluation (Fig. 12). The answers of the participants resemble an equal distribution, even if a complete differentiation is difficult. Nevertheless, it is clear that great importance is attached to validation and that workarounds have now been established to deal with the physicality of mechatronic systems.

5 Discussion of Findings on the Use and Understanding of Prototypes

The results of the study series show that the value of regular and process-accompanying prototyping is recognized and its intention is also understood. One of the initially frequently mentioned hurdles in the use of prototypes, namely the regular generation of functional prototypes, was relativized in practice by significantly expanding the interpretation of what can be used as a prototype. All artefacts that are created during development and allow statements to be made about the functionality of the resulting product that can also be interpreted by the customer are now perceived as prototypes. It is obviously also accepted that prototypes can only depict partial functionalities. It is also recognized and clear that the information content of the prototype fluctuates in different phases of development. In the early phases, greater uncertainties in the description or product functionality must still be taken into account. Here, virtual product artefacts used are certainly perceived as representative prototypes. Therefore, both virtual and real prototypes are used. It is crucial that the information content of these can be categorized.

One assumption from software development is that the entire product is virtually conceptualized and structured as a prototype that can be discussed with the user. These prototypes are completed selectively by the increments. The result is a complete product, but with varying degrees of maturity. This approach proves to be feasible because concept descriptions are purely function-orientated, which means that input/output relationships can also be defined at an early stage. This does not seem realistic for the development of large, complex systems for a number of reasons.

The detailed development of mechatronic systems is also based on functional descriptions. For the identified functions, however, it is necessary to name operating principles which, in turn, are coupled with specific material, energy and information flows in the sense of functional fulfilment, which are then combined to form active structures. This is also associated with specific installation space requirements and requirements for integration into an overall system. Concretization and detailing give rise to parameters, relationships and interfaces that can only be anticipated to a very limited extent.

The development of mechatronic systems is usually asynchronous. This means that not all development tasks are processed simultaneously in the development teams, as the time required for individual development tasks simply varies. In some cases, sequential partial developments are necessary in order to be able to make meaningful decisions for the definition of other, further development tasks. Ultimately, there is a discrepancy in the perception and definition of increments in relation to prototypes. While a series of increments successively complete the prototype in software development (see Fig. 13 left), increments arise depending on the division of tasks, for which it must ultimately be considered how these are to be summarised in order to make certain functionalities tangible in the sense of prototypes (see Fig. 13 right). This requires a dedicated reflection on the division of tasks. If we assume that only partial functions are mapped in the development of large, complex systems, we are ultimately faced with the challenge of structuring development tasks in such a way that prototypes can be derived that actually allow conclusions to be drawn about the system behaviour of the overall system.

Mechatronic systems are characterized by a high degree of functional diversity, not only because solution approaches from different domains are integrated, but also because

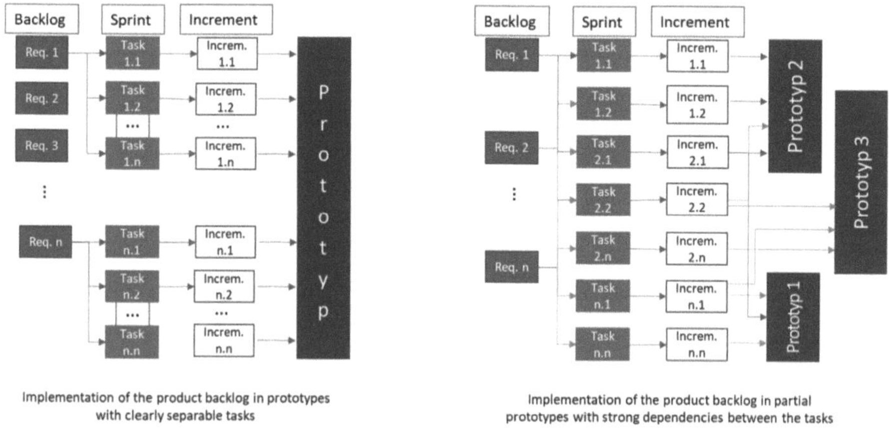

Fig. 13. Possible relationships between increment and prototype depending on the division of tasks

a large number of sub-functions are required to generate a defined overall system behavior. Division of labor is also necessary in order to provide the knowledge required to generate solution approaches for the subtasks. Although agile methods offer the potential to overcome silo thinking, specialization in knowledge will continue to be necessary in order to generate high-quality work results. In vehicle development, for example, specialised knowledge is required for engine or powertrain design, as well as for body construction or interior design. This manifests itself not least in the definition and interpretation of prototypes. The results of the study clearly show that vertical prototypes, i.e. prototypes that only represent components of the entire technical system, are widely accepted and used.

To summarize, it is clear from the studies on understanding and prototypes that individual elements of agile development must not be viewed in isolation from one another. Prototypes and their use and design cannot be interpreted in isolation from the conglomerate of activities, roles and artefacts within the process. This is made clear not least by the diffuse use of increment and prototype. If we look at the process model for agile working described in Sect. 2, sprint planning is based on the product backlog. This is accompanied by a division of tasks. This in turn means that the various tasks are taken up and processed within a sprint. As part of the sprint planning, a "Definition of Done" is defined in the sense of a target plan, which leads to an increment after successful completion of the task and is transferred to a prototype for validation. This makes it clear that the design, utilization and interpretation of prototypes is not only determined by the type of task division, but also by the development phase in which one finds oneself. This shows that in order to use and adapt agile methods effectively, it is necessary to explain elements of agile development, such as the prototype, in terms of their significance from means-ends relationships and/or supplementary cause-and-effect analyses.

6 Summary and Outlook

To summarize, it can be said that physicality alone is not the challenge in the agile development of mechatronic systems. Above all, the complexity of mechatronic systems, their functional diversity and the high level of functional integration appear challenging, as this is accompanied by a necessary division of labor for which agile methods are not initially designed. Kruchten describes this as the so-called "bitter spot conditions" (Kruchten 2013) (see also Sect. 2). In practice, so-called scaling approaches are available to overcome the challenges associated with development in distributed teams (Dingsoyr et al. 2018), which are intended to help support inter-team cooperation in addition to inner-team cooperation. The importance and utilization of scaling approaches has already been demonstrated by the 2023 study. It remains to be seen how the means-purpose relationships must be interpreted and designed here in order to be able to bring the effects of agile working, namely transparency, flexibility and self-organization, to bear across the board. This in turn requires a deeper understanding of the mechanisms of action in agile working and a clear description of the significance and benefits of the individual elements in agile approaches.

References

Anderson, D.J.: Kanban: Successful Evolutionary Change for Your Technology Business. Blue Hole Press (2010)

Atzberger, A.: Agility in Mechatronics unveiled, thesis (Doctoral), Universität der Bundeswehr München (2021)

Atzberger, A., Paetzold, K.: Current challenges of agile hardware development: what are still the pain points nowadays? In: Proceedings of the Design Society: International Conference on Engineering Design, vol. 1, no. 1, pp. 2209–2218. Cambridge University Press, July 2019

Beck, K.: Extreme Programming Explained: Embrace Change. Addison-Wesley Professional (2000)

Beck, K., et al.: Manifesto for agile software development (2001)

Boehm, B.W.: A spiral model of software development and enhancement'. Computer **21**(5), 61–72 (1988). https://doi.org/10.1109/2.59

Boyd, J.R.: The essence of winning and losing (1996)

Brugger-Gebhardt, S., Brugger-Gebhardt, S.: Die Rolle der Prozesse im Qualitätsmanagement. Die DIN EN ISO 9001: 2015 verstehen: Die Norm sicher interpretieren und sinnvoll umsetzen, pp. 9–18 (2016)

Budde, R., Kautz, K., Kuhlenkamp, K., Züllighoven, H.: What is prototyping? Inf. Technol. People **6**(2/3), 89–95 (1990)

Conforto, E.C., Salum, F., Amaral, D.C., Da Silva, S.L., De Almeida, L.F.M.: Can agile project management be adopted by industries other than software development? Proj. Manag. J. **45**(3), 21–34 (2014)

Dingsøyr, T., Moe, N.B., Seim, E.A.: Coordinating knowledge work in multiteam programs: findings from a large-scale agile development program. Proj. Manag. J. **49**(6), 64–77 (2018)

Dingsøyr, T., Nerur, S., Balijepally, V., Moe, N.B.: A decade of agile methodologies: towards explaining agile software development. J. Syst. Softw. **85**(6), 1213–1221 (2012)

Dybå, T., Dingsøyr, T.: Empirical studies of agile software development: a systematic review. Inf. Softw. Technol. **50**(9–10), 833–859 (2008). https://doi.org/10.1016/j.infsof.2008.01.006

Ehrlenspiel, K., Meerkamm, H.: Integrierte produktentwicklung: Denkabläufe, methodeneinsatz, zusammenarbeit. Carl Hanser Verlag GmbH Co KG (2013)
Gonen, E.: Tim brown, change by design: how design thinking transforms organizations and inspires innovation (2009). Mark. Glob. Dev. Rev. **4**(2) (2020)
Gustavsson, T.: Benefits of agile project management in a non-software development context - a literature review. In: Project Management Development – Practice and Perspectives, Riga, pp. 114–124 (2016)
Hallgrimsson, B.: Prototyping and Modelmaking for Product Design, Hachette, UK (2012)
Komus, A., et al.: Studie Status Quo (Scaled) Agile 2019/2020. Hochschule Koblenz, Koblenz (2020)
Kossiakoff, A., Biemer, S.M., Seymour, S.J., Flanigan, D.A.: Systems Engineering Principles and Practice. Wiley, Hoboken (2020)
Kruchten, P.: Contextualizing agile software development. J. Softw. Evol. Process **25**(4), 351–361 (2013). https://doi.org/10.1002/smr.572
Agile PEP-Minds n.n (2023). https://www.agile-pep-minds.de/. Accessed 20 Feb 2024
Ovesen, N.: The challenges of becoming agile: implementing and conducting scrum in integrated product development (2012)
Paetzold, K.: Data and information flow design in product development. In: Krause, D., Heyden, E. (eds.) Design Methodology for Future Products, pp. 201–218. Springer, Cham (2022). https://doi.org/10.1007/978-3-030-78368-6_11
Punkka, T.: "Agile Hardware and Co-Design", Embedded Systems Conference, No. ESC-3008 (2012)
Rößler, L., Gericke, K.: Analysing paradigms for managing product development: conventional, agile and hybrid approaches. Proc. Des. Soc. **2**, 263–272 (2022). https://doi.org/10.1017/pds.2022.28
Schrof, J., Paetzold, K.: Agile Produktentwicklung in einer zunehmend dynamischen Automobilwirtschaft: Potentiale und Grenzen, in 41. Internat. Wiener Motorensymposium, Wien, pp. 362–376 (2020)
Stelzmann, E.S.: Agile Systems Engineering - Eine Methodik zum besseren Umgang mit Veränderungen bei der Entwicklung komplexer Systeme, thesis (Doctoral), Technische Universität Graz (2011)
Sutherland, J.: Scrum: The Art of Doing Twice the Work in Half of the Time. Random House Business Books (2014)
Sutherland, J., Schwaber, K.: The Scrum Guide - The Definitive Guide to Scrum: The Rules of the Game (2020). https://scrumguides.org/
VDI: 2221: VDI-Richtlinie 2221 Blatt 1: Entwicklung technischer Produkte und Systeme - Modell der Produktentwicklung. Beuth Verlag GmbH, Berlin (2019)
Walker, W.E., et al.: Defining uncertainty: a conceptual basis for uncertainty management in model-based decision support. Integr. Assess. **4**(1), 5–17 (2003)
Weiss, S., Paetzold-Byhain, K., Michalides, M., Pendzik, M., Scharold, F., Stoiber, L.: Agile Entwicklung physischer Produkte 2023. Studienserie verfügbar unter (2023). https://tu-dresden.de/ing/maschinenwesen/imm/vpe/forschung/studie-agil. Accessed 20 Feb 2024

Evolution Towards Smart PSS Development – Industry Insights

Yevgeni Paliyenko(✉), Daniel Roth, and Matthias Kreimeyer

University of Stuttgart, Stuttgart, Germany
yevgeni.paliyenko@iktd.uni-stuttgart.de

Abstract. The extensive digitalization and interconnectivity of products and services, has transformed the industrial landscape. In this context, data-driven business models, such as smart Product-Service Systems (smart PSS), have emerged as key enablers of value creation and data-based offerings throughout the product's lifecycle. Despite these opportunities, the development and integration of smart PSS remain a challenge for enterprises.

This paper addresses the gap between academic research and industrial practices through an explorative study of German SMEs, aiming to map and analyze their current development practices. Our research focuses on identifying challenges, opportunities for improvement, and necessary changes within existing processes towards smart PSS development. The study consists of expert workshops, which were conducted in 5 companies with a total of 22 experts from relevant domains.

The findings reveal that development practices in these enterprises are predominantly informal and shaped by legacy systems, which, while adaptable, hinder the full realization of smart PSS potential. Our study highlights the critical need for enterprises to adopt a more systematic approach to development processes. By effectively integrating product and service development with data-centric tools and methodologies. Additionally, the research provides actionable insights for practitioners to expand and enhance existing competencies towards successful smart PSS development.

1 Introduction and Motivation

The rapid evolution of digital technologies and increasing interconnectivity have fundamentally transformed the industrial landscape over the past few decades. This transformation, often referred to as Industry 4.0, is characterized by the pervasive digitalization and networking of products and services. As a result, there has been a shift in how products are conceived, developed, and utilized. In this new era, data and information have emerged as central pillars of value creation, not only optimizing existing offerings but also enabling the creation of entirely new functionalities and business models (Abramovici 2018; Otto et al. 2018).

At the core of this industrial revolution is the emergence of digital business models that leverage extensive data use to generate value (Chowdhury et al. 2018). Within this context, smart Product-Service Systems (smart PSS) have gained popularity as they integrate intelligent products with digital services into comprehensive solutions tailored to

K. Gericke et al. (Eds.): DSC 2022/2023, PTT, pp. 130–148, 2026.
https://doi.org/10.1007/978-3-032-16278-6_9

customer needs (Zheng et al. 2021; Valencia et al. 2015), These systems create additional value for all participants in the value network by offering flexibility, adaptability, and continuous improvement throughout the entire product lifecycle (Valencia et al. 2015; Porter and Heppelmann 2014).

Despite the opportunities presented by digitalization, the development and implementation of smart PSS pose significant challenges. Companies must manage the complexity inherent in these systems, efficiently process large volumes of data, and integrate diverse components into a cohesive offering (Heuchert et al. 2020; Paliyenko et al. 2022). A deeper analysis and understanding of the development processes and practices are crucial to optimizing the performance of smart PSS and supporting their continuous improvement.

Academic research has long focused on the fundamental questions of how, why, and what - particularly regarding the integration of theoretical concepts into practical applications. In recent years, researchers have explored the challenges of integrating these theoretical constructs into real-world practice, offering insights from various industries and providing valuable perspectives on smart PSS. However, existing research on smart PSS development has primarily focused on production and logistics, with only limited attention given to the product development processes themselves (Toller and Bertoni 2021, Abramovici 2018). Moreover, there is a noticeable gap in the literature regarding industrial (non-academic) perspectives on smart PSS and their development, particularly from the viewpoint of small and medium-sized enterprises (SMEs) (Filho et al. 2017). This gap is significant because SMEs play a crucial role in many economies, and their experiences can provide valuable insights into the practical challenges of developing smart PSS.

To address this gap, we conducted an explorative study involving German SMEs. The main objective was to expand the existing literature by adding a practical perspective on current practices in the development of smart PSS. By capturing, mapping, and analyzing existing development processes, we aimed to identify the potential for improvement and the need for change within these processes. Our research focused on answering key questions, such as how the development processes currently used in German SMEs can be improved to better support smart PSS development, and where changes in these processes are necessary.

Our study aims to capture the current state of development processes for smart PSS in German SMEs and derive actionable insights that can inform the industry. By analyzing these processes, we hope to provide guidelines that can help enterprises better navigate the complexities of smart PSS development. Through this research, we seek to contribute to a deeper understanding of how development processes must evolve to meet the demands of the modern industrial landscape and ensure the successful implementation of smart PSS.

2 Exploration of Industry Practices and Development Processes

The study focused on examining the current development practices and processes within five German SMEs, each representing distinct sectors, varying positions within their value chains, and differing company sizes as shown in Table 1. These companies, traditionally centered around physical goods with non-digital downstream services, were in the process of transforming their portfolios through digitalization, aiming to evolve into smart PSS. This transformation provided an excellent context for understanding the challenges and opportunities associated with such a shift, offering deep insights into the ongoing evolution toward smart PSS development.

The participating companies were at various stages of development toward smart PSS at the time of the workshops. While some were in the early stages of developing smart, complex systems, others had already made significant progress in creating sophisticated products and integrated systems. This diversity in developmental stages provided a rich basis for comparative analysis and helped to identify common challenges and success factors in the transition to smart PSS.

Table 1. Case overview and workshop participants

Enterprise description and employees	Case #1	Case #2	Case #3	Case #4	Case #5
	Product manufacturer of laser technology (<250)	Parts supplier of plastic seals for system solutions (<250)	Hardware start-up for battery systems (<20)	Single-order producer of electric components (<250)	Modular contract development (<50)
PSS design case	Optics for laser machining	Hydraulic valve systems	Uninterruptible power supply	Condition monitoring of a switch cabinet	Industrial Floor for electrostatic discharge
Participants	After-sales manager	Director of digital transformation	CEO	Director of R&D	Project manager
	Product manager	Director of technology and innovation	Product developer	Director of software development	Project manager
	Product manager	IoT developer	Product manager	Software developer	CEO
	Service manager	Service manager		IoT developer	PR manager
		Product developer		Product manager	

To thoroughly investigate these transformation processes, company-specific workshops were conducted. These workshops aimed to explore the companies' current development practices, understand their existing processes, and identify actionable needs for improvement. Participants in these workshops were encouraged to share their employee-specific perceptions, allowing for a comprehensive view of the development processes within each organization.

2.1 Workshop Design and Data Collection

The workshops were structured around several work units, focusing on two primary tasks: 1) collaboratively modelling the current product and service development processes within each company and 2) identifying critical intersections between product and service development. The experts had four hours to complete these tasks, ensuring a focused yet comprehensive exploration of the relevant issues. The workshops were conducted using the digital collaborative platform "Conceptboard," which allowed participants to simultaneously and independently create and modify content. This platform was chosen for its ability to facilitate real-time collaboration, essential for capturing the dynamic and interactive nature of the workshops. All participants were provided with instructions on using the platform, and the supervising researchers were available to offer support for any operational issues that arose.

The workshops were led by two academic moderators and were supported by two additional academic observers from other research institutions. These observers participated silently, later contributing their observations and insights to the evaluation process. The workshops involved a total of 21 experts, each selected for their expertise in smart PSS development as described in Table 1.

Prior to the workshops, instructive presentations and discussions were held to establish a common understanding of the terminology, tasks, and goals. These preparatory activities were essential in aligning the participants' perspectives and ensuring a smooth and effective workshop process. Based on the initial statements from participants, enterprise-specific boards were prepared to facilitate an easy and efficient start to the collaborative activities.

The workshops began with an introductory presentation of existing development models (see Table 2), which provided a theoretical foundation for the discussions. These models included the 3-Cycle Model of Product Creation (Gausemeier 2009), the Integrated Product Creation Model (Albers and Braun 2011), VDI 2206 for the development of mechatronic and cyber-physical systems (VDI 2206, 2021), the, and the model for the development of digital service systems (DIN SPEC 33453, 2019). These models served as a reference point for the systematic approach used in the workshops, which involved first documenting the phases, followed by tasks, associated roles, and finally, the methods and tools employed.

To guide the discussions, company-specific guidelines were developed, drawing inspiration from relevant interview content. These guidelines ensured that the workshops were closely aligned with each company's unique context. A lively exchange of ideas was encouraged, fostering targeted discussions and detailed modeling of the development processes. In addition to the digital documentation captured on Conceptboard, written notes were taken to ensure that both explicit statements and implicit nuances were recorded. This dual documentation approach was critical for drawing insights about unspoken needs and potential areas for process improvement.

Overall, the workshops were designed to address the primary research question of the study: to investigate and understand the current development practices within these enterprises, visualize existing processes, and identify necessary changes for the future. The insights gained from these interactive workshops were integral to the subsequent

Table 2. Overview of presented development models

Model (reference)
VDI 2221- Design of technical products and systems (VDI/VDE 2221, 2019)
IPEMM - Integrated Product Development Process Management Model (Albers 2007)
Spiral Model (Boehm 1986)
DevOps (Erich 2014)
Development of mechatronic and CPS (VDI/VDE 2206, 2021)
W-Modell: Systems engineering in the development of active systems (Nattermann and Anderl 2010)
3-cycle model for product creation (Gausemeier 2013)
Generic PSS development process model (Stark and Müller 2010)
Designing sustainable PSS (Sarancic et al. 2023)
Development of data-based PSS (Aurich 2019)
Societal embedding of sustainable PSS (Ceshin 2014)
Configuring New Business Models for Circular Economy through PSS (Pieroni 2019)
Development of digital service systems (DIN SPEC 33454, 2019)

analysis, providing a solid foundation for understanding the broader implications of smart PSS development in the industry.

2.2 Findings and Analysis of Development Processes in Practice

The subsequent findings are presented as they were originally generated by the workshop participants, with minimal alterations. These modifications were limited to translation adjustments and minor post-processing edits, ensuring that the results accurately reflect the participants' intentions. As a consequence, some development phases or steps may not be distinctly separated or might initially appear misplaced. However, these findings genuinely represent the current practices and processes as they are executed in the respective enterprises.

The visual depictions of the development processes have been relocated to a dedicated chapter at the end of this section. Due to their size and the space they require, this adjustment has been made to enhance the overall reading quality and flow of the document.

2.2.1 Findings: Product Manufacturer of Laser Technology

The first company is one of the SMEs that has established a well-defined and documented development process specifically for product design. The development process (see Fig. 2.2.1) is structured around a stage-gate model, which systematically considers the entire product development cycle, including production and sales. This structured approach ensures that every phase of the product development is carefully managed, from initial concept to final market release.

Before the main development process begins, the company employs a pre-development phase, which is selectively implemented based on the novelty or scale of the project. This phase is critical for ideation and evaluating the feasibility of new ideas, as well as accurately estimating the resources required and identifying potential obstacles. The pre-development phase includes activities such as reviewing the existing product portfolio, generating and assessing new ideas, creating preliminary product data sheets, and developing proof-of-concept models. However, smaller development projects may bypass this pre-development stage and proceed directly to the main development process.

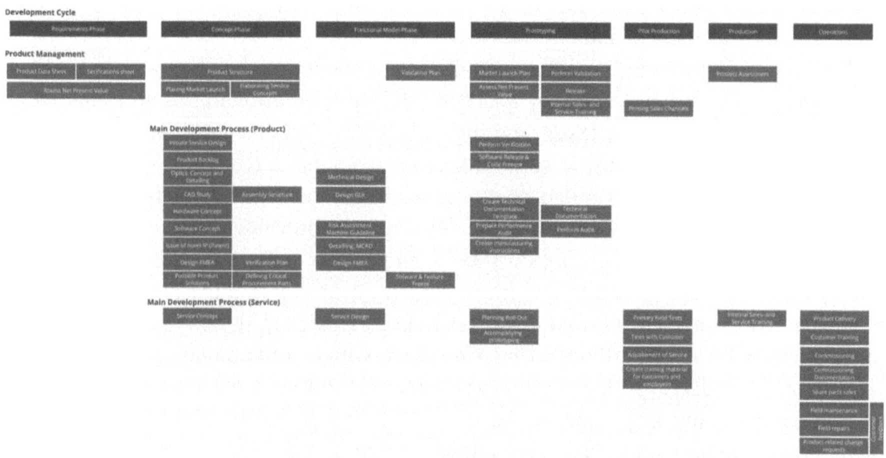

Fig. 1. Development process of case #1

The core development process commences with a rigorous requirements phase, during which the product requirements are thoroughly elicited and defined. This phase involves the creation of a Product Data Sheet and a Specifications Sheet, which serve as foundational documents guiding the subsequent development stages. A critical aspect of this phase is the evaluation of the net present value (NPV) of the project. Only if a positive NPV is confirmed does the project proceed to the next steps, such as concept design.

Upon completion of the product concept, it is handed over to the service development team. The product manager initiates the process of developing service concepts by first exploring potential and necessary services that complement the product. At this stage, the product manager communicates directly with the service developers to outline these service possibilities. Notably, this interaction is limited to a single session, with no further collaboration between the product and service development teams until the product prototype phase. The service development team does not proactively generate service ideas, as their role is reactive rather than proactive at this stage.

Service development is primarily concentrated on traditional, product-related services such as commissioning, maintenance, repair, and spare part supply. There is limited consideration or exploration of more complex or data-driven services, which could offer significant value in a digitally integrated product-service system.

From a technical standpoint, prototyping plays a crucial role in the process, serving as a means to verify and validate the product's design and functionality. Prototyping ensures that the product meets its design specifications and functions as intended before advancing to production. Once a prototype has been successfully validated, the company moves into pilot production, followed by full-scale production. This transition includes field testing and any necessary adjustments based on the test results. Concurrently, training manuals are developed and distributed to both external stakeholders, such as customers, and internal stakeholders, including developers and sales teams. These manuals are essential for ensuring that all parties involved have a comprehensive understanding of the product and its associated processes.

The service department's primary activities are concentrated in the operations phase, which begins with product delivery and commissioning. Following these initial steps, the service department undertakes ongoing tasks such as maintenance and repair, often conducted in the field. To ensure continuous improvement, a feedback loop is in place, allowing after-sales teams to collect customer feedback and issue product-related change requests, which are then fed back into the development process.

This feedback mechanism not only helps in refining existing products but also in identifying areas for innovation and improvement, ensuring that the company remains responsive to customer needs and market demands. However, there is potential for further integration of advanced, data-driven services, which could enhance the company's offering in a rapidly evolving digital landscape. Exploring these opportunities could position the company as a leader in the smart PSS domain, leveraging digitalization to deliver more comprehensive and value-added services alongside their physical products.

2.2.2 Findings: Parts Supplier of Plastic Seals for System Solutions

This company, which specializes in the production of C parts, has established distinct processes for both product and service development, maintaining separate teams for each. In response to the evolving landscape of smart PSS, the company has set up a dedicated service development department. This department has been linked to the product development process (see Fig. 1), although product development still sets the pace and remains the dominant force in the overall development activities. Despite the clear separation, there are numerous defined interfaces and interaction points between product and service development. The service development team, however, plays an increasingly influential role, with multiple points of interaction and involvement in the product development process (Fig. 2).

The company's products are typically simple in design, which is reflected in their lean development process compared to other enterprises. Notably, the product development team does not consider the implementation, market launch, and after-sales activities as part of the "development" process. These responsibilities are handed over to the sales team, which takes charge of these critical activities.

Fig. 2. Development process of case #2

The overarching development process is initiated by the product team, beginning with an ideation phase. This phase leverages various triggers to generate product ideas, which may originate from the company's strategic roadmap, the technology radar, which utilizes expert insights to generate new ideas, or direct customer inputs. Feasible ideas are then transferred to the pre-development portfolio, where they undergo assessment by multiple experts. If the assessment is positive, the idea moves into agile pre-development, where iterative concept development is carried out to explore potential solutions.

The most promising concepts advance to the main product development stage, where a business case is carefully evaluated. The concepts are then further refined, with iterative focus on optimizing design aspects such as shape and material. Upon achieving technical release, the development project is transitioned to the service team.

The service team, in parallel, conducts its own ideation phase early in the process, deriving potential service offerings from historical and current market, customer, and sales data. These offerings are discussed with customers and internally evaluated to build a profitable business case. Once the product is finalized, the service delivery is tested on pilot customers and refined based on the insights gained. This stage provides critical information for value creation within the network, emphasizing the profitability of the product-service bundle. Upon successful development, the company evaluates outsourcing and make-or-buy options to optimize the delivery of the product-service offering.

Despite the defined processes and interactions between the product and service teams, the separate nature of these teams and processes leads to redundancies, particularly in the initial phases of development. This redundancy indicates an area where further integration could streamline development tasks and improve overall efficiency.

2.2.3 Findings: Hardware Start-Up for Battery Systems

This company is the smallest among the SMEs analyzed, yet it stands out as a high-tech producer with a strong focus on innovative, smart products. Despite its extensive experience in developing hardware, software, and mechanical components, the enterprise had never formalized or visualized its development processes prior to the workshop. This lack of formalization reflects a more ad-hoc and flexible approach to development, which has both advantages and limitations.

Development within this company is typically driven by specific, product-related projects, which are initiated in response to various triggers such as direct customer requests, necessary product improvements, or the development of entirely new products. Regardless of the trigger, every new idea undergoes a rigorous evaluation for technical and economic feasibility, as well as an assessment of its novelty. Only after the idea has been examined by experts and a clear product vision has been established does the company launch a formal development project. This selective approach ensures that resources are invested in projects with a high likelihood of success and clear market potential.

The development process (see Fig. 3) itself unfolds in three parallel strands, each dedicated to a core area: mechanics, software, and hardware development. Each strand begins with the domain experts defining the specific requirements for their respective areas. This individual focus allows the company to leverage the specialized knowledge of its team members. However, recognizing the interdependencies among these domains, the company also emphasizes overarching coordination of system requirements to identify mutual dependencies and points of interaction early in the process. This coordination is crucial for ensuring that the various subsystems integrate smoothly into a cohesive final product.

As the development progresses, the domain teams work on their respective subsystems, maintaining bilateral communication to address any adaptation requirements that arise due to dependencies between different components. This continuous coordination helps to mitigate risks associated with subsystem integration and ensures that the final product meets both technical specifications and functional requirements. In parallel, the company maintains close contact with suppliers, manufacturers, and customers throughout the development process. This ongoing communication is vital for validating feasibility, functionality, and compliance with industry standards, particularly given the company's emphasis on product certification. Adhering to relevant directives and guidelines is a priority, and these requirements are carefully translated into system requirements and implemented during development.

Given the small size of the company, decisions regarding outsourcing or make-or-buy strategies are continuously evaluated. This pragmatic approach allows the company to balance its internal capacities with the need to meet development goals efficiently. However, it also reflects a need to be agile and adaptable, given the limited resources available in-house. Interestingly, the company does not consider production and sales as part of the development phases. These functions are seen as distinct from the core development activities, which may limit opportunities for integrating feedback from manufacturing and sales into the development process. Similarly, there is no dedicated service development within the company, and no specialized experts are employed in this

Fig. 3. Development process of case #3

domain. The services offered are based on traditional, straightforward offerings such as installation, maintenance, and repair. This indicates a potential area for growth, particularly if the company aims to enhance its smart product offerings with more sophisticated, data-driven services.

The small number of employees results in developers often taking on multiple roles, which can be both a strength and a challenge. On one hand, this cross-divisional knowledge enables a broader understanding of the product and its requirements, facilitating better integration across different areas of development. On the other hand, it can stretch resources thin and may lead to inefficiencies if not managed carefully. Nevertheless, the small team size allows for continuous, informal exchanges among team members, which fosters a high degree of flexibility. This flexibility is one of the company's key strengths, enabling it to quickly adapt to changing project needs or external demands.

Overall, while the company's development process is highly flexible and tailored to its specific needs, it lacks the formalization and strategic alignment that could help it scale its smart product offerings more effectively. To fully capitalize on its strengths and mitigate its limitations, the company might consider formalizing certain aspects of its development process, especially in integrating service development and production feedback into the development cycle. By doing so, it could enhance its ability to deliver more comprehensive and competitive smart PSS solutions in the future.

2.2.4 Findings: Single-Order Producer of Electric Components

The enterprise under discussion is characterized by its focus on developing highly technological products with a modular product architecture, allowing for significant reuse of prior solutions. This strategic reuse underpins their development process, which is both structured and iterative, enabling the efficient creation of both strategic and customized products.

The development process (see Fig. 4) begins with a discovery phase, which is distinctly segmented into two tracks: strategic products and customized products. Strategic products are conceived based on comprehensive market studies, customer surveys, and targeted ideation sessions. These are products aimed at aligning with broader market trends and emerging customer needs. Customized products, on the other hand, arise from explicit customer requests or are driven by product management's initiatives to improve existing offerings. Regardless of the origin, each product idea is carefully assessed for its compatibility with the company's existing product portfolio.

Fig. 4. Development process of case #4

Once a product idea passes this initial conformity check, the system requirements are derived, and a preliminary, but incomplete, specification data sheet is created. This data sheet serves as the foundation for an internal review of the development project, focusing on key aspects such as general feasibility, cost-effectiveness, and crucial make-or-buy decisions. This review ensures that only viable projects proceed to the next stages of development.

Following this, the development process moves into the concept phase. Here, the requirements and specifications are refined, and a thorough assessment of the technical and time-related feasibility is conducted. During this phase, the requirements for the product can be iteratively developed and adapted by the development team, allowing for flexibility and refinement. Once the functional specification is firmly established, it becomes the baseline for the remainder of the development process, with any subsequent changes requiring a formal change management procedure.

In the concept phase, solution concepts are generated for the product architecture, production processes, and scalability considerations. Concurrently, a development project is formally initiated, with detailed scheduling, budgeting, and project milestones established, each with a stage-gate character to ensure progress is regularly evaluated.

The enterprise's development process is organized into three main development strands: Software, Hardware, and Mechanics. The requirements are divided among these strands, with close attention paid to identifying mutual dependencies and points of interaction between them. In the software development strand, the project is broken down into manageable tasks, which are tackled individually through an iterative development cycle. Each task is solved independently before being integrated into the broader system. For hardware development, the process begins with an evaluation of existing solutions from the company's modular solution kit, with an emphasis on reusing tried and tested PCB layouts where possible. The design team makes necessary adaptations, followed by electrotechnical analyses to ensure functionality. The spatial distribution of electronic components on the board is then finalized, leading to the creation, manufacturing, and preliminary testing of the final PCB layout. The mechanics strand focuses on the development of the housing for the hardware components. This involves computer-aided design and simulation, with particular attention to collision detection and thermal management to ensure the physical integrity and performance of the product.

As soon as the three development strands - software, hardware, and mechanics - have sufficiently matured, a prototype is developed and subjected to rigorous testing. This prototype phase is iterative, often involving the creation of multiple prototypes to identify and resolve any issues that arise. The enterprise emphasizes rapid development and rapid prototyping to accelerate the process, particularly since hardware development typically lags behind software development. To mitigate delays, the teams are encouraged to use standard components and simulations wherever feasible, thereby shortening the time to the physical prototype or final product.

Once the development process is complete, the product undergoes intensive quality assurance, including numerous functional, thermal, and fire load tests. After successfully passing product qualification, the product enters series production and is prepared for market launch. During this phase, the company offers a range of classic services, including commissioning, customer training, and system configuration.

In the operational phase, following the product's sale, the company continues to provide traditional services such as updates, repairs, and customer support. However, these services are not integrated into the development process and are primarily aimed at maintaining the functionality of the system rather than enhancing its value or exploring new revenue streams. This reflects a strong focus on ensuring the reliability and durability of the products but a relatively limited emphasis on leveraging services to generate additional value.

2.2.5 Findings: Modular Contract Development

The enterprise offers a relatively straightforward product from a technological perspective, which results in a development process (see Fig. 5) that adheres closely to classical product development methodologies. The company's development approach is highly customer-centric, focusing on engineering to order. This orientation ensures that the product is developed according to specific customer requirements, making the process extremely tailored but also limiting the exploration of innovative business models, smart features, or additional services.

The enterprise's product structure and business model are closely tied to commissioning and assembly, which are integral components of the value offered to customers. In this context, these phases are considered essential parts of the development process, with significant attention given to them early in the development cycle. This focus begins during production planning, where considerations for proper packaging dimensions and correct assembly are integrated into the design to ensure that the final product meets all customer specifications and logistical requirements.

The development process is organized around projects, each triggered by various internal or external sources, such as sales opportunities, specific customer requests, or marketing initiatives. Once a project is initiated, the development process begins with a requirements phase. This phase involves internal planning, which includes initial sketches, coordination with suppliers, and close collaboration with customers. Additionally, feasibility checks are conducted to ensure that both the development and production processes can meet the outlined requirements.

Following the requirements phase, the process moves into the concept phase, where the design is developed using CAD tools, and technical 2D drawings are generated. This phase is characterized by numerous iterations, involving both the customer and the internal development team, to refine the design. After the customer approves the concept, a final revision is performed to ensure manufacturability.

The subsequent production phase is comprehensive, starting with detailed production planning and including considerations for packaging and assembly. Material procurement follows these plans, and a first product sample is produced and presented to the customer for final approval. Upon receiving the customer's approval, the full-scale manufacturing process is initiated.

Given the size and complexity of the products, both commissioning and assembly are pivotal to the enterprise's value proposition. These activities are not merely logistical tasks but are central services offered by the enterprise, contributing significantly to the final product's value. Despite their importance, other services such as maintenance and repair are considered secondary and do not play a prominent role in the development process. Consequently, these services are not systematically integrated into the early stages of product development, which may limit the long-term service potential and lifecycle value of the products.

This enterprise's development process is deeply rooted in traditional engineering practices with a strong emphasis on meeting specific customer demands through tailored product development. While the process is efficient and customer-focused, the lack of integration of advanced digital features and the minimal emphasis on service-related aspects during development may present challenges as the industry increasingly shifts toward smart and connected products.

Fig. 5. Development process of case #5

3 Insights and Implications for Practitioners

In examining the development practices of the SMEs, it became clear that none of the companies adhere to a single, unified development process. Instead, their processes have evolved organically over time, shaped by the specific needs of their prior portfolios and the cumulative experience of their teams. These organically grown processes often retain remnants of outdated practices, reflecting a legacy of the companies' historical focus rather than a deliberate shift toward modern, theoretically grounded frameworks.

One significant issue observed across these enterprises is their reluctance to critically evaluate and challenge their existing development structures. When these companies initiated smart PSS development, they predominantly relied on their traditional, established processes, even though these were originally designed for a different era and different types of products. This reliance on legacy processes often hampers the ability to fully exploit the potential of smart PSS, as the old frameworks are not always equipped to handle the complexities and integration requirements that modern, data-driven, and service-oriented products demand.

The lack of experimentation and innovation in process development is particularly concerning. None of the enterprises attempted to pilot smart PSS projects with a clean slate approach - an opportunity to experiment with untied hands, free from the constraints of legacy processes. By sticking to familiar methods, these companies missed the chance to explore novel development approaches that could have been better suited to the unique demands of smart PSS development. This conservative approach limits the potential for discovering more efficient, agile, and innovative processes that could better align with the dynamic nature of smart PSS development.

Moreover, the existing practices in these enterprises are not based on a coherent, single development model. Instead, they represent a patchwork of various established development frameworks, selectively combining elements from multiple sources. These practices incorporate best practices and lessons learned from past projects, which, while valuable, can further entrench legacy thinking and prevent a holistic re-evaluation of what the process should be in the context of smart PSS. In addition, the reliance on patchwork integration of various development models can create confusion and reduce the effectiveness of the development process. When processes are patched together from multiple sources without a clear, overarching strategy, it can lead to misalignment between different teams and phases of development. This fragmentation makes it difficult to achieve the kind of seamless integration that smart PSS requires, where product and service development need to be closely coordinated and mutually reinforcing.

Furthermore, the use of legacy processes can inhibit the company's ability to fully leverage new technologies and methodologies. For example, the traditional stage-gate models often used in these enterprises may not be flexible enough to accommodate the iterative, agile methodologies that are more appropriate for developing integrated product-service solutions. Without the adoption of new, more adaptable processes, companies risk falling behind competitors who are more willing to innovate and experiment with their development approaches.

To make the development process more suitable for smart PSS development, enterprises must undergo a significant transformation in how they approach product and service design, integrating these processes into a seamless, cohesive workflow. This evolution requires not only the adoption of new methodologies but also a cultural and operational shift within the organization, ensuring that the development process is more integrated, responsive, and agile.

Central to this transformation is the need for closer integration of product related development activities and service related activities. Enterprises should move away from treating these processes as separate, sequential tasks and instead adopt a concurrent engineering approach. This means that both product and service teams work together from the initial concept phase, ensuring that services are designed in tandem with products. Such integration leads to more cohesive and comprehensive offerings that are better aligned with customer needs. To facilitate this, enterprises must create cross-functional teams that include product developers, service designers, software engineers, data scientists, and customer specialists. These teams should be empowered to collaborate closely, ensuring that both the product and its associated services are fully integrated and optimized for a smart PSS from the start.

Digitalization and data-driven strategies are also critical components of this evolution. IoT technologies and data analytics should be embedded in the product design phase, enabling the creation of products that can collect, transmit, and analyze data in real-time. This capability opens up new possibilities for data-driven services, such as predictive maintenance and performance optimization. Enterprises should invest in developing digital twins that allow for real-time monitoring, simulation, and testing of products in a virtual environment. By doing so, companies can significantly reduce development time and costs while improving product quality and reliability.

To further support this data-driven approach, enterprises should incorporate big data and AI capabilities into their development processes. This involves not only capturing and analyzing large volumes of data but also using AI to predict market trends, optimize performance, and personalize services. Establishing data pipelines that continuously feed real-time data from operations into the development process allows enterprises to adapt and evolve their PSS offerings based on customer usage patterns and implicit feedback, thus continuously improve them over time.

Adopting agile and iterative development practices is another essential step in evolving the development process. Traditional stage-gate models, while effective in certain contexts, may be too rigid for the dynamic nature of smart PSS development. Instead, enterprises should implement agile methodologies that allow for iterative development and continuous feedback loops. Agile practices enable faster iterations, quicker prototyping, and more frequent updates, which are critical for responding to rapidly changing customer needs and technological advancements. The implementation of continuous integration and continuous deployment practices further accelerates the development and deployment of both products and services, ensuring that the smart PSS can evolve dynamically in response to new data, user feedback, or technological innovations.

In addition, enterprises should design their PSS offerings to be modular and scalable. This means developing products and services that can be easily adapted or upgraded over time to meet changing customer needs or to enter new markets. Modular design principles allow companies to offer flexible and customizable solutions that cater to a wide range of applications, while scalability ensures that the smart PSS can grow and evolve as the business expands.

Enhanced collaboration and communication are vital for the success of this integrated approach. Enterprises should leverage integrated communication platforms that facilitate real-time collaboration, version control, and documentation. These platforms should be accessible to all team members and stakeholders, ensuring that everyone involved in the development process has access to the latest information and can contribute effectively. Additionally, companies should establish regular co-creation workshops, customer feedback sessions, and beta testing phases to involve stakeholders - especially customers - throughout the development process. This active involvement ensures that the final PSS offering is closely aligned with customer needs and expectations.

Customer experience and value proposition must be central throughout the development process. This involves thoroughly understanding customer needs, pain points, and expectations at every stage of the product and service lifecycle. User experience design principles should be integrated into both product and service development, ensuring that the final offering delivers maximum value to the customer. Enterprises should continuously refine their value proposition based on market research, customer feedback, and emerging trends, ensuring that it remains relevant and compelling.

A holistic lifecycle perspective is also crucial. Enterprises should not only focus on the initial design and production phases but also consider the entire lifecycle of the product and service, including end-of-life management. This approach involves designing products and services with sustainability in mind, such as minimizing environmental impact, optimizing resource use, and enabling circular economy practices like remanufacturing and refurbishing.

By implementing these process-specific changes, enterprises can transform their development processes to be more suitable for smart PSS development. This transformation requires a cultural shift toward greater integration, collaboration, and agility, supported by digitalization and data-driven strategies. By doing so, enterprises will be better suited to develop and offer smart PSS solutions that meet the complex demands of customers.

4 Conclusion and Outlook

The comprehensive study has provided a thorough examination of the current development practices and challenges faced by German SMEs in the context of smart PSS development. Our analysis revealed that the development processes within these enterprises are largely informal and have evolved organically, often carrying forward remnants of legacy systems and approaches. While this flexibility allows for some degree of adaptability, it also hampers the ability to fully harness the potential of smart PSS, particularly in terms of integrating digital and data-driven components. The reliance on ad-hoc processes, coupled with a lack of formalization and strategic alignment, presents significant obstacles for these enterprises as they strive to innovate and remain competitive in an increasingly digitalized market.

The findings of this study highlight the critical importance of systematically addressing the complexity inherent in smart PSS development. The effective integration of product and service development processes, alongside the adoption of digital tools and methodologies, is essential for these companies to maintain their competitive edge. Moreover, our research emphasizes the need for a more formalized and methodical approach to development, one that can manage the multifaceted nature of smart PSS by considering the entire product lifecycle - from initial design of the business case through to post-sale services during operations - ensuring all aspects of the system are seamlessly addressed and optimized.

Looking forward, there is a clear necessity for further research and practical development to deepen our understanding of the specific roles, responsibilities, methods, and tools that underpin effective smart PSS development. By delving deeper into these process descriptions, we can create a more granular understanding of how smart PSS development can be efficiently managed and executed within different organizational contexts. This deeper insight will enable enterprises to better align their resources and capabilities with the demands of smart PSS development.

Additionally, there is significant value in developing a meta development process for smart PSS, drawing from the insights gained through this study and future research. This meta-process would serve as a structured framework, encapsulating the most important lessons learned and best practices, thereby providing SMEs with a clear reference point as they navigate the transformation process of integrating product and service development with digital technologies.

Finally, it will be crucial to revisit the participating enterprises at a future time to assess how their development processes have evolved. By examining the changes implemented and their outcomes, we can derive valuable industry-based lessons that will inform and guide other companies seeking to enhance their smart PSS development practices. This longitudinal analysis will offer critical insights into the effectiveness of various strategies and approaches, ultimately contributing to a more robust understanding of how to successfully manage the development of smart PSS in the dynamic landscape of modern industry.

Acknowledgements. This work has been supported by the German Federal Ministry of Education and Research through the research project "bi.smart" (grant no. 02J19B043).

References

Abramovici, M., Gebus, P. and Savarino, P.: Engineering smarter Produkte und Services Plattform Industrie 4.0 STUDIE. acatech – Deutsche Akademie der Technikwissenschaften, Munich, Germany (2018)

Albers, A. and Meboldt, M.: Integrated Product Development Process Management Model, based on system engineering and systematic problem solving. International Conference on Engineering Design. Cité des sciences et de L'industrie, Paris, France (2007)

Aurich, J.C., Koch, W., Kölsch, P., Herder. C.: Entwicklung datenbasierter Produkt-Service Systeme - Ein Ansatz zur Realisierung verfügbarkeitsorientierter Geschäftsmodelle. Springer Vieweg Berlin, Heidelberg, Germany (2019)

Boehm, B.W., TRW Defense System Group 1988Boehm, B.W., TRW Defense System Group: A Spiral Model of Software Development and Enhancement. Computer, Institute of Electrical and Electronics Engineers, vol. 21, no. 5, pp. 61–72 (1988)

Ceschin, F.: Sustainable product-service systems: between strategic design and transition studies. Springer International Publishing (2014)

Chowdhury, S., Haftor, D., Pashkevich, N.: Smart Product-Service Systems (Smart PSS) in industrial firms: a literature review. Special Issue: 11th CIRP Conference on Industrial Product-Service Systems, Procedia CIPR, vol. 73, pp. 26–31 (2018)

Erich, F., Amrit, C.: Report: DevOps Literature Review. Technical Report, University of Twente (2014)

Filho, M.F., Liao, Y., Loures, E.R., Junior, O.C.: Self-Aware Smart Products: Systematic Literature Review, Conceptual Design and Prototype Implementation, Procedia Manufacturing, Vol. 11 Special Issue: 27th International Conference on Flexible Automation and Intelligent Manufacturing, pp. 1471–1480 (2017)

Gausemeier, J.: Strategische Planung und integrative Entwicklung der technischen Systeme von morgen. Schriftenreihe der Nordrhein-Westfälische Akademie der Wissenschaften, Band 42, Verlag Ferdinand Schöningh, Paderborn, Germany (2014)

Heuchert, M., Verhoeven, Y., Cordes, A.K., Becker, J.: Smart Service Systems in Manufacturing: An Investigation of Theory and Practice. Proceedings of the 53rd Hawaii International Conference on System Sciences, pp. 1686–1695 (2020)

Heuermann, A., Olschewski, D., Wiesner, S.A.: DIN SPEC 33453:2019-09 Development of smart services, Beuth Verlag GmbH (2019)

Nattermann, R., Anderl, R.: Approach for a data-management-system and a proceeding-model for the development of adaptronic systems. In: ASME International Mechanical Engineering Congress & Exposition, pp. 1–9. ASME, Vancouver, Canada (2010)

Otto, B., Hompel, M., Wrobel, S.: Industrial Data Space - Referenzarchitektur für die Digitalisierung der Wirtschaft. In: Neugebauer, R. (ed.) Digitalisierung – Schlüsseltechnologien für Wirtschaft & Gesellschaft, pp. 113–133. Springer, München (2018)

Paliyenko, Y., Tüzün, G.-J., Roth, D., Kreimeyer, M.: Inquiry and analysis of challenges in the development of smart product-service systems, international design conference - DESIGN 2022, pp. 1935–1944. Cambridge University Press, Cambridge, USA (2022)

Pieroni, M.P.P., McAloone, T.C., Pigosso, D.C.A.: Configuring new business models for circular economy through product-service systems. Sustainability, MDPI **11**(13), 3727 (2019)

Porter, M.E., Heppelmann, J.E.: How smart, connected products are transforming competition. Harvard Business Review (2014)

Sarancic, D., Pigosso, D.C.A., Pezzotta, G., Pirola, F., McAloone, T.C.: esigning sustainable product-service systems: A generic provess model for the early stages, vol. 36, pp. 397–414. Sustainable Production and Consumption, Elsevier Ltd (2023)

Stark, R.G., Müller, P.: A generic PSS development process model based on theory and an empirical study, International Design Conference - Design 2010. DESIGN 2010, pp. 361–370. Dubrovnik, Croatia (2010)

Toller, C.N.K., Bertoni, M.: The research domain of product-service systems and voice of the customer: a systematic mapping. In: Proceedings of the International Conference on Engineering Design (ICED21), ICED21 1, Gothenburg, Sweden, 16–20 August 2021 (2021)

Valencia, A., Mugge, R., Schoormans, J.P.L., Schifferstein, H.N.J.: The design of smart product-service systems (PSSs): an exploration of design characteristics. Int. J. Des. **9**(1), 13–28 (2015)

Verein Deutscher Ingenieure. VDI 2221, Entwicklung technischer Produkte und Systeme - Modell der Produktentwicklung. VDI, Duesseldorf, Germany (2019)

Verein Deutscher Ingenieure. VDI/VDE 2206, Development of mechatronic and cyber-physical systems. Duesseldorf, Germany (2021)

Zheng, P., Chen, C.H., Wang, Z.: Smart product-service systems. Elsevier, Nanyang, Singapore (2021)

Knowledge Management with MBSE in Battery Development at AVL

Andreas Braun(✉), Ibtihal Badi, Alexander Gelner, and Lena Höllein

AVL, Ingolstadt, Germany
Andreas.Braun@avl.com

Abstract. The following chapter illustrates recent efforts of AVL – a leading engineering service provider – to address Knowledge Management (KM) in industrial practice. Due to its intricate complexity, battery development has been selected as a use case to investigate the chosen approach. Batteries as a black box may seem simple, but the view inside battery development reveals a combination of technological and organizational complexities, emphasizing the technological and organizational challenges involved. Battery development encompasses various disciplines, including chemistry, thermodynamics, electronics, software engineering, and mechanical engineering. Organizational complexities arise from the need of the involved stakeholders to deal with evolving requirements, varying global standards, and different use cases for batteries. Effective project communication and data management are crucial, in particular if multiple stakeholders from various disciplines are involved. Model-Based Systems Engineering (MBSE) is deployed in AVL to manage these complexities. This divides the battery system into subsystems and focuses on 17 key functionalities that are common for every battery system. Variant management is used to handle product variability and to identify synergy potentials between projects. Finally, KM efforts aim to capture project management knowledge in ongoing projects and to provide gained insights to future projects, again. The case study shows that manual maintenance of knowledge resources yields only limited success. The chapter hence concludes with a deep dive into the potentials of integrating artificial intelligence (AI) to enhance KM and engineering processes.

1 Introduction

"The nations of the world know that the complete defossilization of transportation, our industry, actually our entire living environment by 2050, will be one of the most important undertakings in our lifetime." AVL Home – E-Mobility (AVL 2022).

Many companies aim their efforts to similar goals. In this chapter, we are illustrating how engineering methodology, in particular MBSE, can help to pursue this target effectively and efficiently. Since defossilization is still relatively new, it is important to ensure effective processes to capture knowledge and experiences in a systematic way. The development of modern propulsion systems is beset by tight timeframes, a multitude of varying constraints, and extensive requirements, necessitating the collaboration of experts from diverse disciplines. Effective communication, information exchange,

K. Gericke et al. (Eds.): DSC 2022/2023, PTT, pp. 149–181, 2026.
https://doi.org/10.1007/978-3-032-16278-6_10

and documentation play vital roles in this process. As a result, AVL embraced the adoption of MBSE already in 2007, which has since proven to be successful in our customer projects.

In the following paragraphs, the MBSE methodology deployed is introduced, and examples of the models and development artifacts created using MBSE are provided. One of the key advantages highlighted in this chapter is the ability to structure, record, and share knowledge in terms of externalized information and experiential insights. Together with the means of variant management and ultimately with the help of AI technology, a comprehensive approach to enhance effectiveness and efficiency of industrial engineering can be achieved – as illustrated by the example of battery development.

2 Literature

In this section, the basic concepts of the developed KM approach based on MBSE are defined.

2.1 Knowledge Management

KM makes use of certain tools, methods and procedures to efficiently manage resources and information within institutions such as commercial organizations' (Sandhu & Tripathy 2018). KM wants to create value for the company through implementing a process that supports effective sharing and reuse of knowledge (Sandhu & Tripathy 2018). The knowledge stairs by North are the most popular definition of knowledge in literature. The stairs show the development of signs to data to information. According to this, knowledge is information in a context. North points out that KM should not be treated simply as a must-have for every company (North, 2016). Instead, it is more important to focus on the actual benefit knowledge-orientated work has for the whole company. Figure 1 depicts North's knowledge stairs, which figuratively show that KM can be used to manage data or strategically to create an advantage on the market.

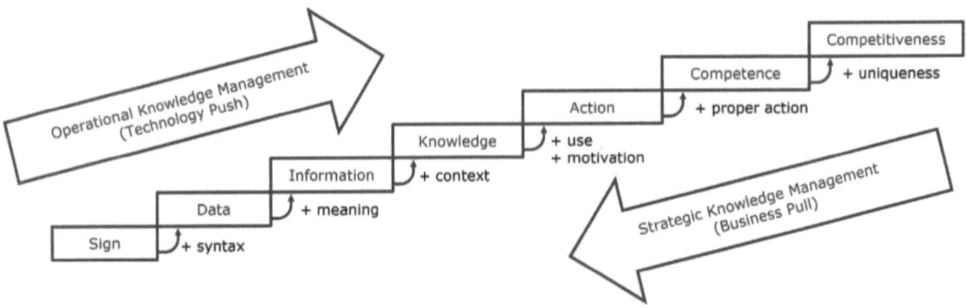

Fig. 1. Knowledge Stairs by North, own representation according to (North 2016)

Al-Kadi (Al-Kadi 2003) outlines the essential properties of a knowledge management system (KMS) as follows:

- Scalable and Extensible: The database must be capable of adapting to accommodate the growing number of artifacts and users.
- Secure: Ensuring the protection of company and customer knowledge is imperative.
- Collaborative: The KMS should be accessible to a wide range of users, promoting knowledge exchange and teamwork.
- Allow for complex queries: The system should enable users to effectively find the information they seek through intricate search queries.
- Flexible: The system needs to have the capacity to handle all types of knowledge utilized and requested by the company.
- Heuristic: The system should learn from users' interactions and needs, enhancing the quality of search results over time.
- Suggestive: The system should provide user recommendations and display relevant results, even those the users were previously unaware of.

2.2 Systems Engineering and MBSE

Systems Engineering (SE) describes a transdisciplinary approach to manage systems of every level of complexity (INCOSE 2019). The idea of SE can be used in many contexts detached of the structure of the system that needs to be enabled. The system can include cyber- or physical components, as well as interaction with the environment or humans (Yang, et al. 2020). The SE approach embraces methods, processes, and tools to successfully master such systems. The methods include aspects known from project management tasks, like requirements and risk management, leadership, and handling of human factors, as well as technical factors (INCOSE 2019). These consider hardware and software components, their integration and verification and validation (INCOSE 2019). SE takes the whole life cycle of a system into account and improves the overall system thinking of the developer and manager.

A complex system cannot be understood as the sum of its parts (Simon 1969). The composition of the parts features complex relationships and is much more than the sum of its parts, thus it needs to be controlled methodologically. Model-based Systems Engineering (MBSE) is one part of SE and offers a methodical approach to organize, coordinate, and collaborate in complex projects, facilitating the integration of requirements, design, analysis, and validation from the project's inception and throughout its lifecycle. The MBSE approach describes the systems as a model. In contrast to a document-based development process the idea of MBSE is that structured graphics are clearer to understand than many lines of text. Especially for people from different knowledge fields, figures and diagrams are easier to understand than text with specialist terms. The system model is created in formalized language such as Systems Modeling Language (SysML) or Unified Modeling Language (UML) to define a common understanding of the meaning of the blocks and connections in the system model (Anderson 2014). The model displays not only the system components and functions but also supports the requirement management, design decisions, system analysis and the optimization, verification, and validation processes (Anderson 2014).

Furthermore, MBSE approach serves also as a benchmark for stakeholders involved in the product development process, from conceptual design to verification and validation, effectively managing the risks arising from increasing product complexity and tight schedules.

2.3 Collaboration

Collaboration involves communication, coordination, and cooperation between members of a team, working together to create and share knowledge.

2.3.1 The Intersection of KM and Collaboration

KM inherently involves collaboration, and various technologies can assist in supporting these practices. Tools for collaborative KM enable individuals to share documents, provide comments, participate in discussions, create schematic diagrams, and more. These tools are valuable in facilitating organizational learning. They incorporate diverse functions, including email, newsgroups/bulletin boards, task lists, document management with features like check-in/check-out systems and version control, as well as workflow management. Figure 2 depicts a generic framework of collaborative KM.

Furthermore, KM and collaboration exhibit a complementary relationship, as asserted by (Nunamaker, et al. 2001). (Qureshi, et al. 2004) contend that these two domains share common purposes and practices, illustrating their mutual interdependence. This interdependence is elucidated through the mapping of collaboration technologies to KM activities: 'Technologies for knowledge management may enable improved capture and conveyance of understanding that might otherwise be inaccessible in isolated pockets; technologies collaboration may enable communication and reasoning among people who use knowledge to create value' (Qureshi, et al. 2004).

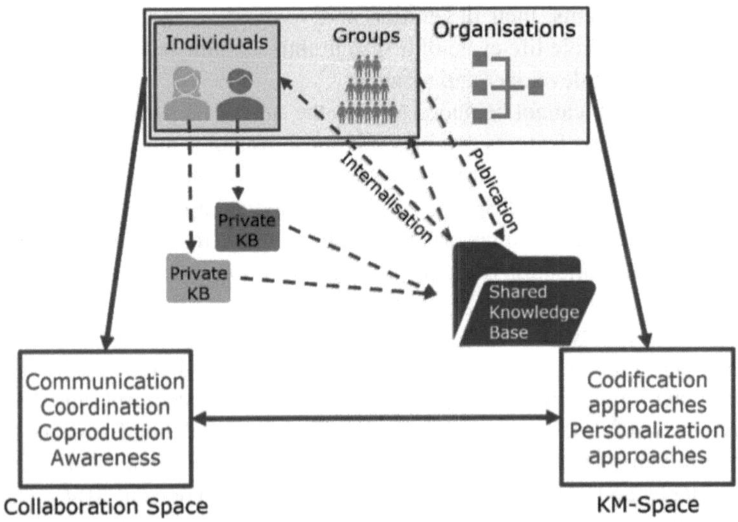

Fig. 2. Generic framework of collaborative KM, own representation according to (Bellamine & Ghézala 2012)

2.3.2 Collaborative Knowledge Management (CKM)

The CKM framework unites a KMS and a Collaborative Information System (CIS) to represent an organization's Intellectual Bandwidth (IB). IB includes both explicit, codified data stored in KMS and CIS, and implicit, non-codified information existing outside these systems (Nunamaker, et al. 2001). As technology and organizational experience advance, the hierarchies of KMS and CIS converge. Together, they optimize and amplify IB to help stakeholders address issues. Figure 3 illustrates an integrated framework where the capabilities of KMS and CIS synergize to determine the accessible IB. Collaborative technologies are positioned within the framework based on their comprehension and collaborative functionalities.

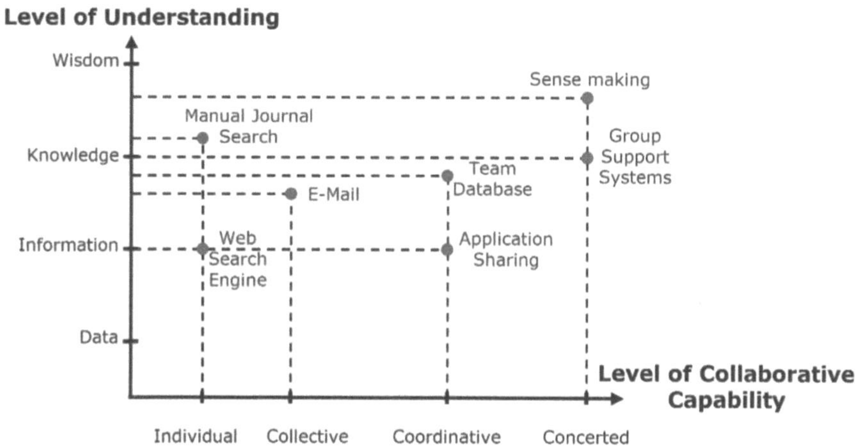

Fig. 3. Integrated Framework of CKM, own representation according to (Nunamaker, et al. 2001)

Figure 4 demonstrates that employing KMS and CIS together leads to a greater quantity of IB within the organization, and the magnitude of this impact is shaped by the size of the organizational group.

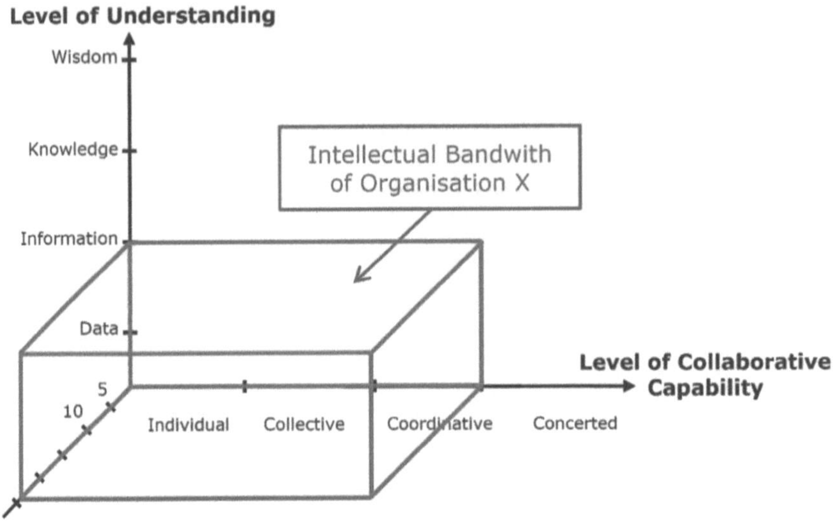

Fig. 4. Intellectual Bandwidth as the product of KM System, Collaborative Information System and Group Size, own representation according to (Nunamaker, et al. 2001).

To increase the IB in a defined team, a method to not only collect, but to coordinate and to concrete the information or knowledge needs to be defined. The more users are intended to participate, the more benefit is to be expected from the method. In order to identify necessary prerequisites of a company-wide implementation and not least also with the aim of identifying limits regarding usability and data volume, a comprehensive approach is being investigated by AVL – in a pilot phase with a focus on the area of battery development.

3 Methodology

To investigate a comprehensive approach, a series of studies has been carried out from 2016 to 2023 and research is still continuing. The hypothesis is: Especially with a combination of MBSE, variant management methods and a KM framework, effective support of engineers in industrial practice can be achieved. The following paragraphs give an overview of the most relevant studies that have been carried out – and why.

The methodology containing process, method and tool which AVL uses for MBSE is outlined in Sect. 4. From 2006 it has been implemented in the area of powertrain engineering at AVL. From 2014 it has been rolled out also for the area of battery development. Detailed descriptions can be found in (Hennige & Braun 2015) and (Braun & Maletz 2015).

After clarifying the general methodology for MBSE in battery development, another series of studies has been conducted with a focus on organizational aspects of the application of such an approach in a company environment. Using methods that have been derived from Lean Management, problems such as time-consuming information and knowledge search have been identified. The studies were conducted in a representative target group of individuals that are working in battery development projects on daily basis. With consecutive surveys and interviews, typical challenges have been derived and quantified in terms of frequency and time effort using the method of a process failure-mode and effects analysis (pFMEA). During the study, not only room for improvement but also opportunities for optimization within projects could be identified. (Müller, et al. 2017).

This has been adopted by (Braun & Löchl 2017) when a new battery development team has been founded. A lean methodology has been introduced which had the intention to realize as many of the identified chances for more effective and efficient work as possible. At the same time as many as possible of the identified problems with complicated methodologies have been tried to avoid. For example, the methodology for the new team has been implemented in the standard Microsoft Office environment. Consequently, users could apply the method of mapping requirements to tasks, functions and physical embodiment directly in their daily work routines without having to be trained in expert authoring tools beforehand (which is often necessary, if specific MBSE tools are applied). The methodology has been used in all projects of the newly founded team for over six years. During that time, many observations could be made in terms of action research – leading to continuous adaptions and improvements of the methodology. This includes the implementation in the tool environment (a Microsoft SharePoint-based online solution, which has recently been transferred into a company cloud) and also supporting assets such as the instruction resources for new employees. Further insights could be gained when the approach has been rolled out to other teams in the department. The researchers learned that not only the technological capabilities of the toolchain are a major success factor, but also organizational aspects such as management support and continuous communication and feedback loops with the users. All findings enabled the observed group of engineers and their managers to minimize inefficiencies and other risks inherent in complex projects. (Einkopf & Braun 2017).

Rolling out the approach to more users and more projects, the already identified challenges of keeping databases up to date manually became obvious more and more. With the aim of elaborating on ideas how to improve this, further methods have been developed. In a publicly funded research project on the one hand and in a research program in cooperation with academic partners on the other hand, the structure of the KM artifacts and a method to cope with variants have been investigated in detail. Thanks to the continued support of the management, it was possible to introduce parts of the methods in customer projects in parallel to the research activities. This allowed the researchers to test ideas, discuss potential and to get feedback directly from the target group. It also proved to be very helpful that most members of the target group were already part of previous studies. With the experience from the first research experiments, they were able to contribute even more effectively and helped to bring the approach to a new level.

Based on the results, an ontology has been conceptualized and validated accompanied by a series of expert interviews during a three-semester research Master program. The ontology comprises metadata which has been defined specifically to enhance the KMS with regard to user-friendliness and added value. Artifacts and their relationships are elaborated. With this, the KMS itself could be integrated into the work structures of the employees even better to increase the usage of the system. The foundation for automatization of KM work was given. At the same time, significant improvements in terms of development quality and time to market have been confirmed by the customers. (Novak, et al. 2021b).

In 2023, the usefulness of variant management for derivative development of batteries was inspected as a further addition to the approach. It has been conceptualized by the research team and was elaborated in a Master thesis that was part of the research project MBPLE4Mobility. The work involved analyzing an actual project, coupled with documenting experience related to the modeling tool and process. (Badi, et al. 2024a, b) Furthermore, the opportunities of AI as a technology to support the application of the KM and MBSE approaches have been investigated. Building on the preceding master theses, an internal project was conducted to facilitate an understanding of the possible roles that AI might play in the interactions between knowledge workers and the established approaches of KM and MBSE. (Badi, et al. 2024a, b).

4 Application and Evaluation of the Approach in Industrial Practice

This chapter presents development of KM and collaboration methods at the AVL battery department. It reflects the challenges and discusses the solutions based on the general methods of SE at AVL.

4.1 Setting of the Case Study: Organization Chart

AVL Germany manages three battery teams situated in four locations across Germany as well as two further teams in Hungary and Istanbul. These teams handle diverse customer projects, with comparable tasks. The projects primarily concentrate on the upper segment of the V-model, encompassing project management, requirements engineering, system architecture, as well as verification and validation. Within the framework of our customer projects, AVL assumes complete autonomy in undertaking engineering development tasks based on the customers' requirements and specifications.

Figure 5 shows the structure of the High Voltage Energy Systems Team within AVL Deutschland GmbH:

Fig. 5. AVL Germany Organizational Chart

The global battery competence team consists of more than 600 battery engineers. The collaboration between the teams globally is mainly organized through the central team in the headquarters. To improve the collaboration between the three German teams, different collaboration and KM methods were implemented.

4.2 Systems Engineering at AVL

In 2020, AVL was part of a best practice evaluation organized by the RWTH Aachen (CSE Aachen GmbH 2020). The evaluation discussed the implementation of SE and MBSE methods. AVL was rated to be part of the five best companies in German speaking countries (CSE Aachen GmbH 2020).

The first usages of MBSE approach go back to 2007 at AVL (Denger 2014). AVL developed a method called Systems Engineering for Powertrain Engineering Application (SEPA). SEPA defines the use of a SysML model as a basic component in every project. They offer every new project a generic model which then will be tailored for the project situation. The specific project changes will be added to the copy of the generic model. If the project team creates an additional perspective, which could be useful in further projects as well, the new content will be integrated in a generic version through a change process in the generic model. This process guarantees the reuse of knowledge and lessons learned. The component battery can be found as a part of the powertrain in the vehicle. The SEPA method defines system levels which are ordered top down from vehicle to the vehicle system powertrain to the element battery. Further the battery has for example the component module with the subcomponent cell. Figure 6 shows the system levels.

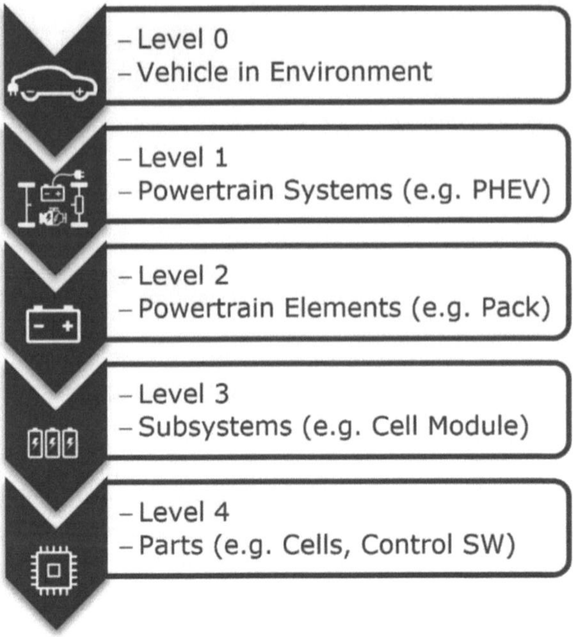

Fig. 6. AVL Level structure, own representation according to (Maletz, et al. 2021)

The SEPA method includes AVL-specific guidelines for requirements engineering and modeling, e.g., it defines standardized boilerplates for functional and non-functional requirement texts, or it defines properties of model elements, so-called stereotypes, for specific kinds of proxy ports in the modeling. The standardized vocabulary for the naming of signals and ports, clear color coding, etc. helps the engineer to understand the diagrams faster. Imagine every project has its own colors and vocabulary. If an engineer switched from one project to another, the onboarding would take longer to understand the project-specific method. The regulations that are made in the SEPA method support efficient collaborative work. The method is applied in AVL engineering projects all over the world and is the basis for a common understanding and the collaboration of different teams.

The SEPA approach at AVL refers to the Requirement view, Functional view, Logical view, and Physical view (RFLP) concept: The requirements are managed in a requirements management tool which is linked to the functions and the logical view modeled in SysML. The physical view is demonstrated in engineering tools, for example CAD modeling. (Maletz 2021) In a development project, the RFLP artifacts are created iteratively as information about the system under development emerges during the process. Using the illustration of a V-Model, activities of integration, verification and validation can be depicted as corresponding levels of the RFLP views on the right side of the illustration (Fig. 7).

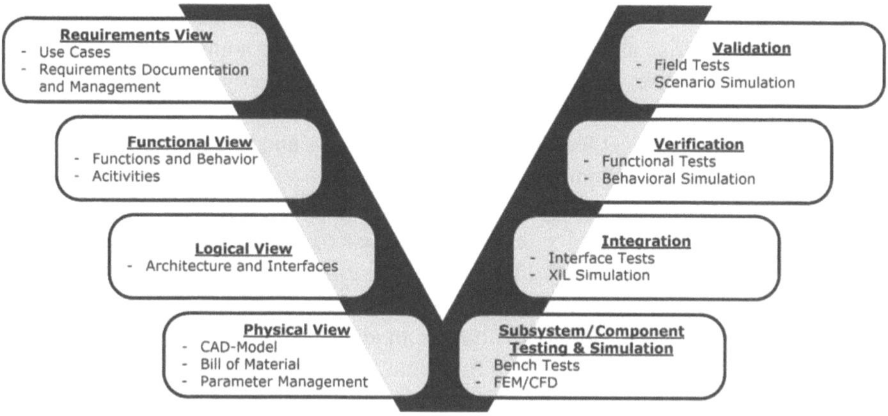

Fig. 7. RFLP concept, depicted in the V-Model, own representation according to (Kleiner & Kramer 2013)

The AVL MBSE follows the principle of hierarchical decomposition and divides the vehicle in systems and subsystems. Recognizing the fact that any truly complex system is not fully hierarchically decomposable, exceptions are possible, for example the thermal system is described as distributed system in every level. Fig. 8 explains the level structure at AVL. The battery is part of the powertrain and starts at level 2 from where it is "broken down" into details.

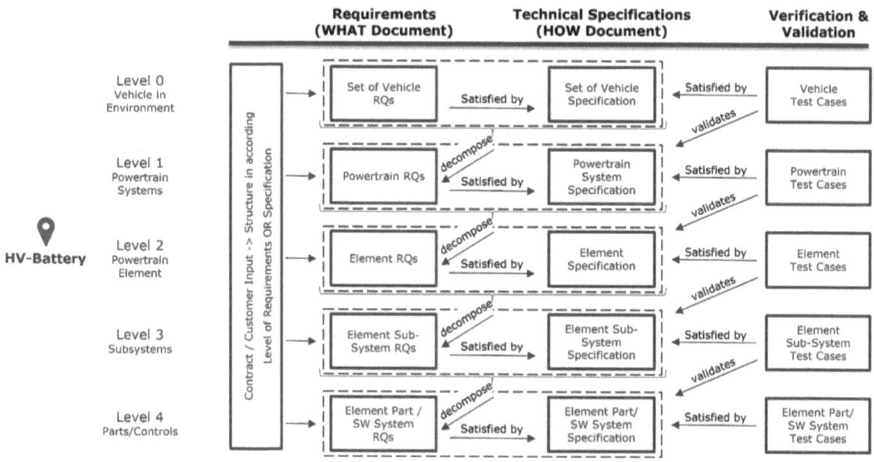

Fig. 8. Zick-Zack Model AVL, own representation according to (Maletz 2021)

4.3 Battery Development – from a Technical Point of View

In the area of battery development, AVL offers different engineering services supporting OEMs from research and feasibility studies to serial development, including simulation and testing solutions. Projects can be very different in scope: Some projects may "only" be one simulation which is carried out by a single person in alignment with the customer. Other projects comprise all activities from concept design to homologation and series production support. Here, several dozen engineers and part project managers from all parts of the world would be involved and need to be coordinated. In some projects, AVL engineers work closely together with customer engineers and people from third parties such as test institutes or sub-contractors. In other projects – so-called "turnkey projects" – AVL works independently most time and interaction with the customer is limited only to the relevant project milestones for review and release of work products. Consequently, the requirements for an approach that is intended to support the knowledge workers are very different from project to project. Still, with their CKM approach, AVL aims at providing know-how to the people involved from all cumulated project experience.

From a technical perspective and at first sight, a battery appears to be quite simple: a box with some interfaces that can store and provide energy. A closer look reveals the complexity of this black box. First, there is the technological complexity: besides electric engineering other multiple disciplines work together in the development. Figure 9 illustrates the disciplines and examples where to find their contributions in a battery pack. Questions concerning chemistry can be found most prominently in the central element of the pack: the cells. These cells need to be cooled, so thermodynamics is relevant as well, and they must be protected from damage and deformation by a mechanical housing. The operating strategy is managed by the battery management system (BMS) in interaction with manifold sensors and actors. Therefore, electrics, electronics and software development knowledge are needed, as well.

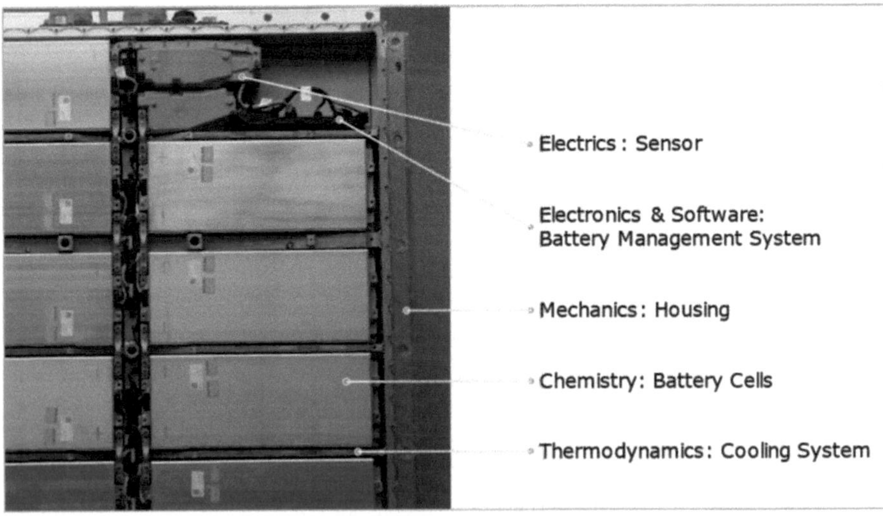

Fig. 9. Volkswagen ID.3 First (AVL Series Battery Benchmark)

So, the battery is a complex system and the experience of the people that are involved in battery engineering is typically limited to their previous project involvements and thus relies only on already seen technologies. Now, the behavior of a battery can be completely different with the use of a different cell chemistry. Another example for complexity is different use cases: Traction batteries and stationary batteries need to fulfil a whole set of different requirements. Management of these requirements and their fulfilment by different part systems needs to be managed comprehensively and with particular regard to the respective use cases at hand.

In the case of traction batteries, it is not uncommon that the requirements evolve or change during the project duration. Further, there are varying standards for batteries all over the world. The standards define for example safety or sustainability aspects for a specific market or region. Therefore, a reliable requirement engineering and change management is needed in the projects. Lessons learned need to be recorded and – most challengingly – be provided in another project at another time, again.

The key focus from customer perspective is the mitigation between performance and costs. In addition to that, the OEM focuses on ideally zero failure in a quantity of million batteries and a lifetime of ten years and more. To achieve this, a conscientious development and testing procedure is indispensable, and a clear development method is needed to ensure a high-quality development process.

Due to the multiple stakeholders of various disciplines which are involved, project communication is a key element in the development method. A common wording and understanding needs to be established in the project team and the interfaces between the disciplines and people need to be defined. Further, a strategy for project data management needs to support the development process for all disciplines from requirements engineering to test management.

As a service provider, AVL has the responsibility to complete the projects it is assigned to, efficiently, sustainably, and conscientiously. Transparent processes and a systematic approach are indispensable for this work.

For instance, handling the complex variabilities in batteries during the development phase is a multifaceted challenge that demands a meticulous approach. To navigate the intricate landscape of the different battery chemistries, collaborating with multiple battery manufacturers, in addition to rigorous testing and validation procedures. All those aspects must be established to account for different solution principles in the product family, ensuring that the final product meets the desired performance and safety standards. That way, battery developers effectively manage the intricate web of factors that influence battery performance, ultimately leading to successful and robust battery solutions.

4.4 Battery Development – MBSE and KM

MBSE constitutes an approach designed to facilitate the development of complex systems by formalizing system artifacts and the development process using models. MBSE models offer comprehensive structural formalisms for system developers and capturing a vast amount of domain-specific technical knowledge. What is not traditionally included is the project management knowledge, e.g., contact persons, abbreviations, templates for the development tasks. This missing information, which is relevant to fulfil the tasks, besides the technical know-how, presents a significant challenge.

In 2016, challenges and potential benefits that could be achieved if advanced work strategies and KM would be implemented efficiently have been identified by means of a comprehensive pFMEA. (Müller, et al. 2017) Time intensive processes, for example searching for the right contact person or specific information, or waiting for access rights, etc. have been analyzed. Further, not reusing documents costs double work for the projects, too. Compare Fig. 3 for information sources and their classification according to literature.

Having understood the challenges, solutions have been developed. To share knowledge more effectively, a clustering of typical activities of the stakeholders in battery development has been developed. A hierarchic coding system for knowledge artifacts has been introduced. It is a combination of the process step, the system level and the task create a classification key which can be added to every artifact as a unique code. Based on these unique classification keys, all the relevant knowledge for a single task can be labelled and found. This method is in theory very efficient and scalable, given that the unified task codes are the same in every project worldwide. (Braun & Löchl 2017).

The unique code is an effective way to organize knowledge among users; but one challenge for efficient usage are the users themselves. An implementation which strongly depends on the user's initiative and willingness to add the unique code manually and to enact the defined guidelines for KM and collaboration, can be problematic when rolled out. A second challenge are the gaps between the tools the classification key is used in. The code is used in different daily process steps, for example for clustering tasks, classifying templates or booking of working hours while giving a reference to the corresponding tasks where the effort occurred. All those things are done with different tools, and these tools are not yet integrated. This is a gap which makes it difficult to analyze the data and to provide a holistic search functionality. While the classification key proved valuable to bridge this gap in principle, there is still the question of how to improve efficiency for the users.

As a basis for a better search process, a knowledge ontology about the context of the respective engineering tasks has been developed. Artifacts from different sources should be included in one shared metadata ontology. Taking the example of a knowledge graph, the nodes are the knowledge artifacts and the single context elements, such as metadata, e.g. "created by" or "user role". The edges connect one artifact with its specific context elements and its related artifacts. Figure 10 shows the artifact nodes with pictograms in the yellow circles and the context nodes in green circles (C1-C4). See (Badi, et al. 2024a, b).

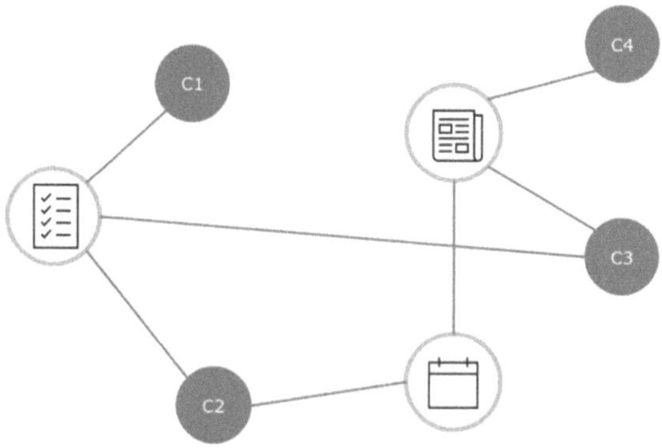

Fig. 10. Knowledge Ontology Example

On the one hand, the use of an ontology improves searching and recommendation. On the other hand, adding knowledge and keeping it updated is a manual process. This takes a lot of time and documentation effort which hinders method ergonomics and thus user acceptance. Apart from that, manual processes are subject to mistakes.

So, in recent work, the use of AI to understand artifacts and give them the right metadata has been investigated. Metadata they can be added automatically to the ontology and the users are supported in the knowledge storing and finding tasks without having to deal with classification schemes or other additional work themselves – it is merely done for them in the background, while they are doing the actual knowledge work. New information can be added to the KMS with the help of AI technologies avoiding human factors for potential errors such as typos. At the same time, user-friendliness is significantly enhanced as close to no manual efforts are required to fill the knowledge base. Finally, AI is assumed to have the potential to facilitate knowledge retrieval from the KMS in different use cases similar to chatbots based on Large Language Models which is subject of ongoing research at the time this chapter is written. (Badi, et al. 2024a, b).

4.4.1 MBSE in Battery Development

In battery development, the battery pack is the system of interest. At this point, in accordance with the AVL MBSE approach, the battery pack belongs to level 2, and the entire so-called RFLP concept is applicable: requirements, functional models, logical architectures, physical implementation.

The model-based battery development program MoBat addresses the enhancements of the MBSE approach in the battery development. The battery is modeled not only as black box, but as a specific subsystem (Braun, et al. 2021). The battery development process is characterized by technical and organizational complexity, such as target conflicts, for example cost vs. performance. The SE methods for process design and the use of MBSE allow controllability and handling of the complexity. (Novak, et al. 2021b).

AVL defines 17 functionalities of the battery system – listed in Table 1 – that must be fulfilled within an automotive application (Novak, et al. 2021b). They take the whole lifetime of the battery into account. Nonetheless, they are important for the production and safe operation of a battery system throughout the lifecycle, like design for manufacturing or homologation and transport. Also, reuse, refurbishment and recycling aspects are considered, to increase the sustainability of the battery systems in the future.

Table 1. AVL definition of a battery's functionalities according to (Novak, et al. 2021b)

ID	Functionality
1	Energy storage and power
2	Thermal conditioning
3	BMS control
4	Geometrical integration
5	Lifetime and reliability
6	Tightness
7	Mechanical robustness
8	HV EE
9	LV EE
10	Electromagnetic compatibility (EMC)
11	Climatic robustness
12	Corrosion and chemical media resistance
13	Labels and markings documentation
14	Safety and abuse
15	Homologation and transport
16	Design for manufacturing (DfM), Design for assembly (DfA)
17	Design for service (DfS), recycling

All the 17 functionalities have their relevance, but the prioritization is different for different customers or use cases. This results in different priorities of objectives during the development and leads to different products. AVL analyses batteries in benchmarking studies and compares them. Figure 11 shows a rating based on factors analysis of three different batteries in different target categories which reflect the quality in which the battery functionalities have been fulfilled.

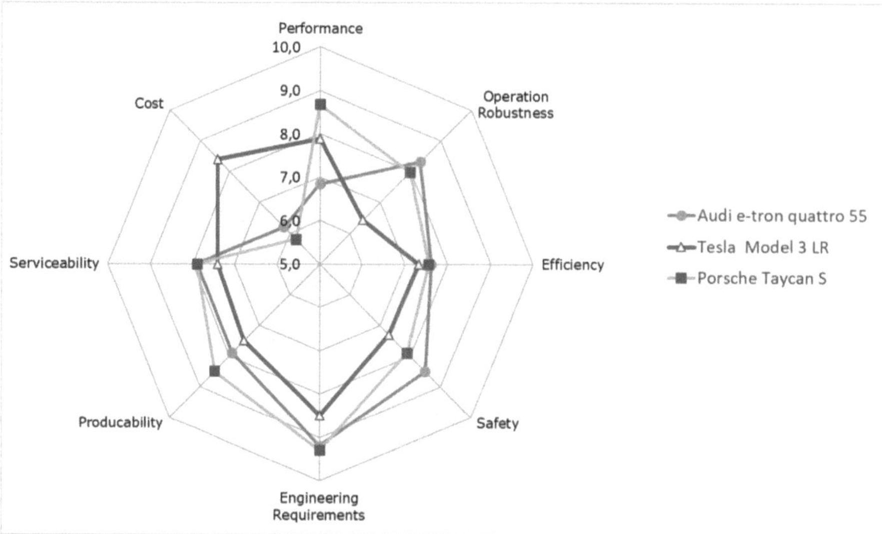

Fig. 11. Targets of Batteries, examples from AVL's benchmarking of purchased cars from series production

Not all objectives can have the highest priority. Typical target conflicts in the battery development result from technical feasibility alone or in combination with economic reasons, for example:

- Energy vs. Weight and Volume
- Power Density vs. Safety and Durability
- Performance vs. Cost: High-performing batteries with advanced materials can be expensive to manufacture. Balancing cost-effectiveness with meeting performance requirements is a continual dilemma for manufacturers and consumers.
- Energy Density vs. Cycle Life: High energy density batteries store more energy, making them attractive for electric vehicles. However, they tend to have a shorter cycle life, meaning they degrade faster over time, which can be a problem for long-lasting applications.
- Charging Speed vs. Battery Lifespan: Rapid charging is convenient, but it generates heat that can accelerate battery degradation. Balancing fast charging with preserving the battery's lifespan is a constant challenge.

To find the right prioritization for the customer and a suitable solution, these conflicts of targets need to be solved. This can be done with respect to the customers' use cases for the battery system and the legislative and standards situation in the target markets.

MBSE diagrams allow visualization and provide a common ground for discussions with the customer and among the designers. Hence, already in 2006, AVL's management introduced MBSE as mandatory approach for all projects where AVL is responsible for the contractual targets. In some projects, AVL is asked to transfer this approach to the customer's engineering teams. For practical application by the different disciplines and stakeholders, AVL's MBSE approach is completed with a common taxonomy. An excerpt can be seen in Fig. 12. This use case diagram gives an overview of the different targets in automotive development.

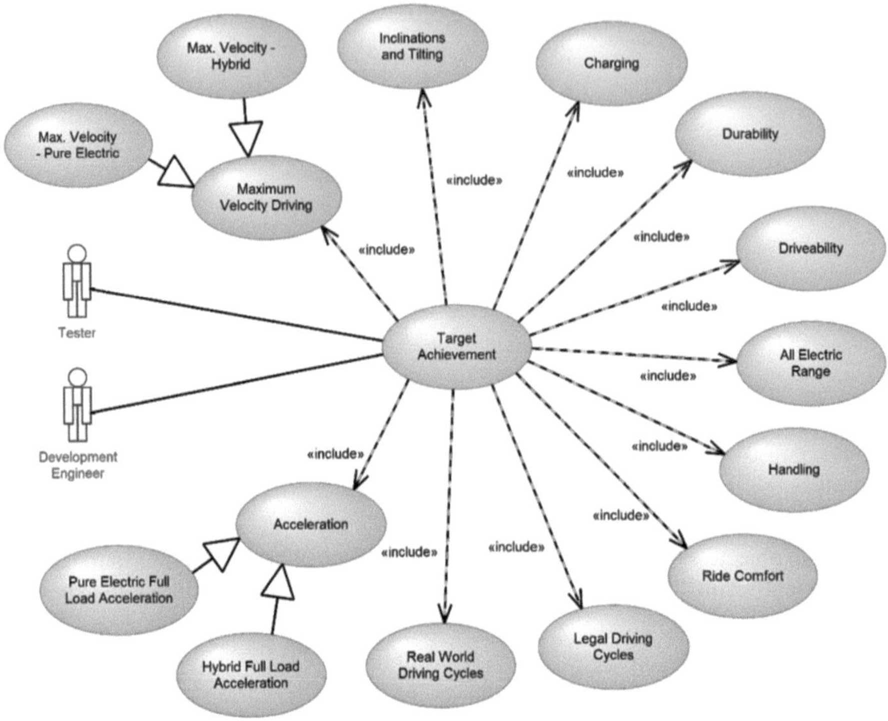

Fig. 12. Target related Use Cases

The subsystems of the battery, shown in Fig. 13, are divided based into the disciplines involved in the development, mentioned above:

- cells, the chemistry system, are ordered in modules or modern cell-to-x systems,
- the housing is the mechanical subsystem (together with other holder structures),
- the thermal system including condensate management, etc.,
- the high voltage (HV) system,
- the BMS and low voltage (LV) system,
- the labels, due to the importance of safety and warning in the HV-Battery development.

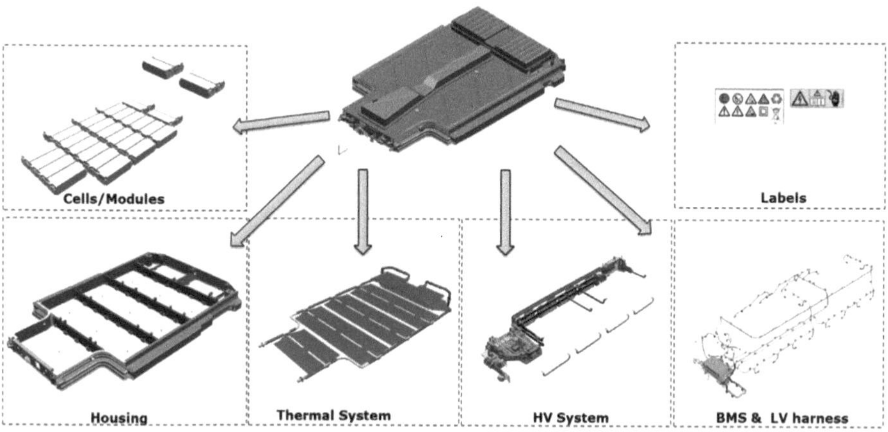

Fig. 13. The Subsystems of a Battery

The battery and its subsystems are modeled with the language SysML and the AVL modeling method. Figure 14 shows the structure of thermal subsystem of the battery. The colors represent different stereotypes.

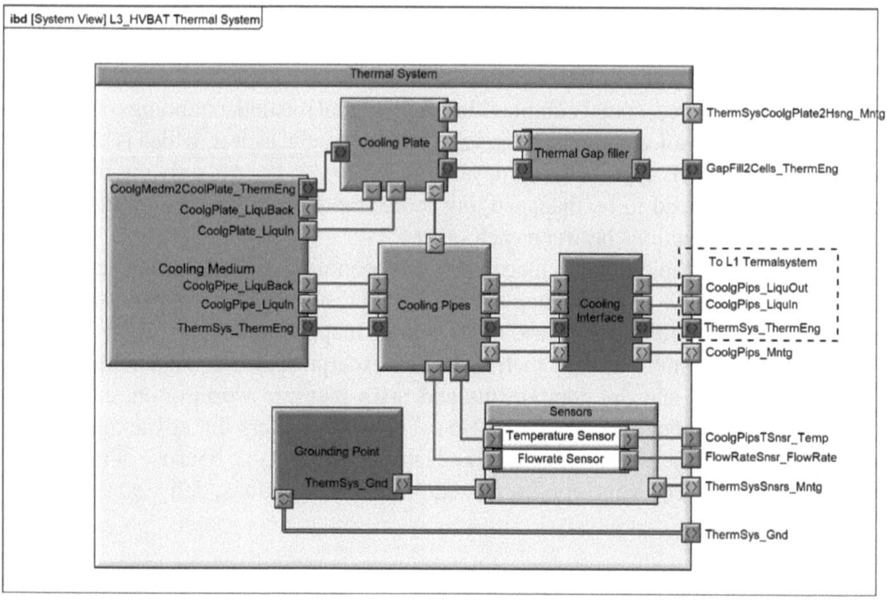

Fig. 14. SysML internal block diagram of the battery thermal subsystem

Further, activity diagrams are used to describe the behavior of the thermal system. Figure 15 shows one such example.

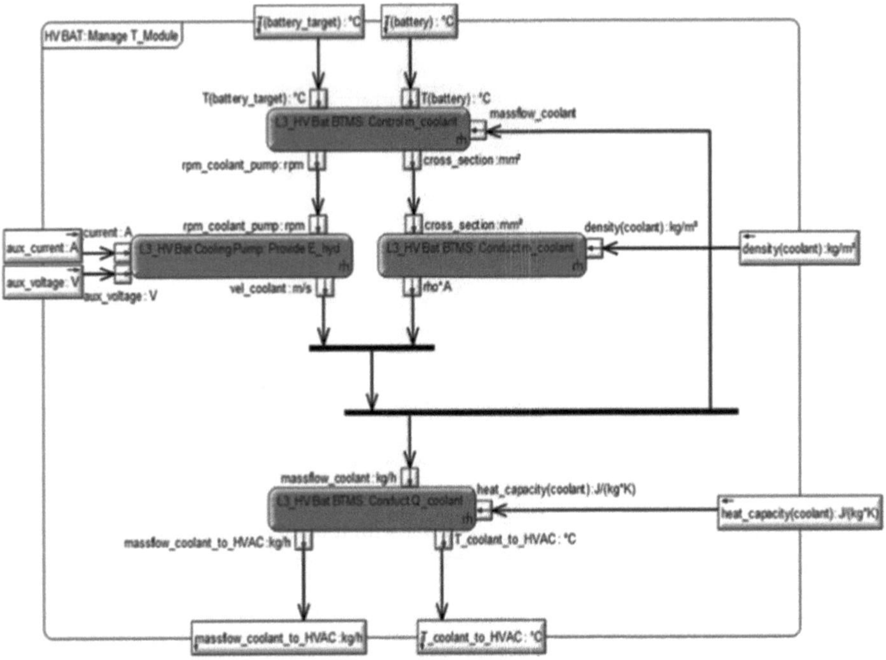

Fig. 15. Example diagram representing functional dependencies (Braun 2021)

As previously noted, social complexities such as goal misunderstandings or miscommunication among project team members are another crucial aspect, which is intended to be met by using the SE approach. In a development project, it is not only the subsystems and functions that need to be managed, but more importantly the team members with their tasks and interactions between each other.

Therefore, a multiple-domain matrix (MDM) according to (Lindemann, et al. 2009) is used to cluster these project items, for example the stakeholder mapped to tasks, the components mapped to functions and the functions mapped to targets. Figure 16 shows a small part of the multi domain matrix. In this excerpt, it shows stakeholders, e.g. a validation engineer, and the components and tasks they are working on, e.g. battery housing in a mechanical test (Minke 2015). The MDM offers the advantage that the indirect relationship between component and task is determinable because of the context via the stakeholder. The same applies for requirements, functions, deliverables, etc.

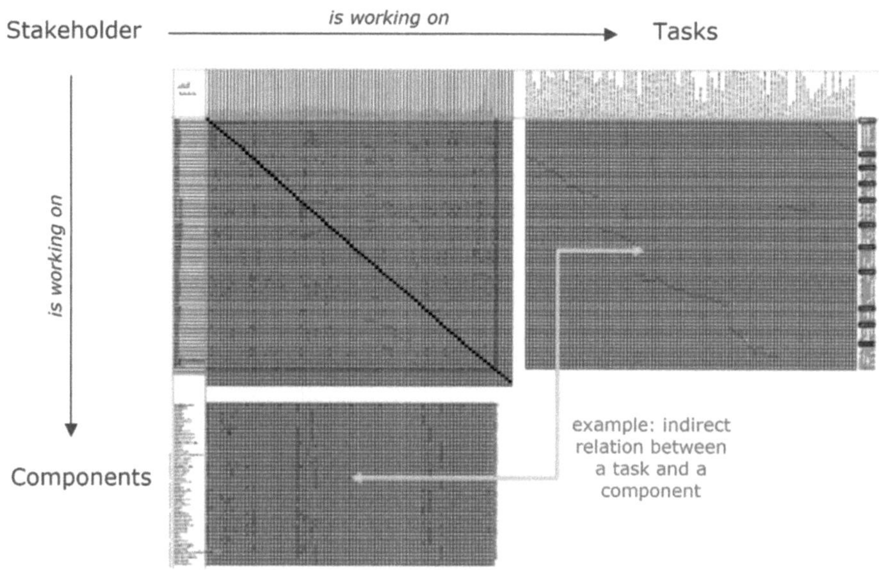

Fig. 16. Part of the multiple-domain matrix, blue arrow shows exemplary dependency, own representation according to (Minke 2015).

Thinking in the categories of the multi domain matrix helps the project members understand the project and to succeed. A prime example for the benefit of this approach has been published by AVL together with Audi in a "Success Story" in 2022 (AVL 2022): A derivate of Audi e-tron has been developed and validated. Using the MBSE approach helped speed up and enhancing quality.

4.4.2 MBSE and Variant Management

In AVL's globally distributed customer and project landscape, variability of products and changes of targets or boundary conditions at the beginning of, or even during the phases of product development, contribute to the complexity of development projects. In current efforts, the "traditional" MBSE approach has been extended to facilitate variant management. This enables the definition and management of variants of products and the artifacts to describe and engineer them.

One aim of variant management is to give an overview of existing variants and their possible combinations of the product. A prominent example in the sector of automotive development is the engineering of derivates. For example, the development of modern automotive batteries is characterized by several variabilities. Variation point can be manufacturer, type, and number of battery cells, different length (wheelbase), two- or four-wheel drive or varying charging standards for instance. Variations can be found in every artifact of the RFLP concept and in every level of detail of the battery system. The overall battery concept is determined on pack level. On subsystem level lay the variation points, for example, in the thermal system e.g., liquid cooled or air cooled, or in the module design. And on component level different cell types, suppliers, chemistries, valves, pumps, fans and other parts and materials influence the overall battery system and ultimately the vehicle attributes and customer experience. OEM-specific norms and market requirements which compile to a big spectrum of standards and corresponding test cases must be managed in a systematic manner: For each variant and in addition to the technical complexity of each individual system itself. All the different kinds of features have an impact on engineering requirements and the required tests.

To manage the correlations of the features of the product, an MBSE method extended by variant management is used to guarantee the traceability between the development artifacts. This method and toolchain have a strong focus on recognizing and reusing existing artifacts for a specific feature from one variant to another or across projects. This enables the development engineers to make use of existing patterns and experiences with those and to give early and knowledge-based feedback to changes and new variants during the development phase. Therefore, variant management increases the understanding of the impact of changes to the existing product, and this can lead to a shorter development time for a new variant of an existing product as highlighted in (Novak, et al. 2021b).

The basis of the variant management approach is a variant description model, also called feature tree, which describes the features that the product can have and the relations between these features. The model can be created out of different perspectives: The customer perspective concentrates on the needs of the consumer and the market. The engineering perspective has a special look at the technical realization of the customer's needs. Figures 17 and 18 show examples of a variant description model from a customer point of view and an engineering point of view.

The customer view focuses for example on specific markets or the driving range needed of the vehicle.

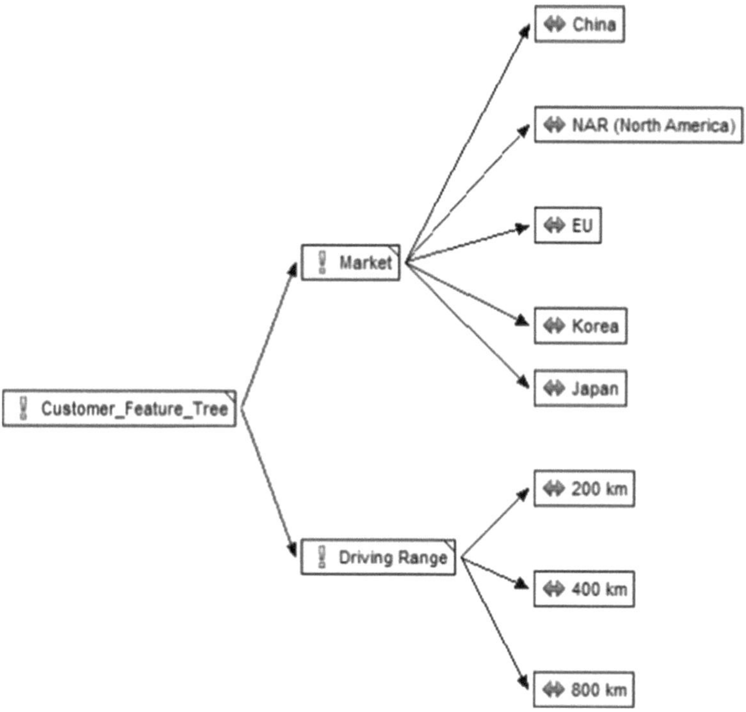

Fig. 17. Variability Description - Customer View

The engineering view distinguishes, for example between different charging systems and different sizes of the battery.

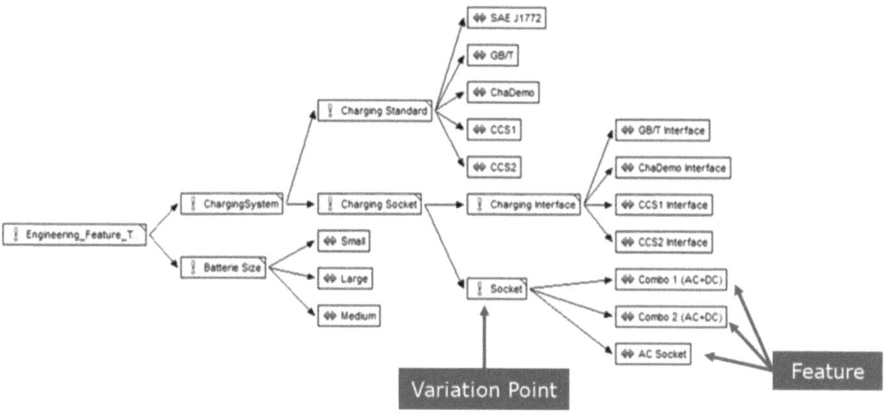

Fig. 18. Variability Description - Engineering View

A feature tree is structured into variation points and the features themselves. The features and variation points can be mandatory items, options, or alternatives. The tree can be enhanced with every relationship between the features which are needed. Further, the different variant description models can be linked to each other. Figure 19 shows the links from the customer model to the engineering model. The charging concept differs based on the market. The linking between the variability description models can visualize the correlation between market and charging concept.

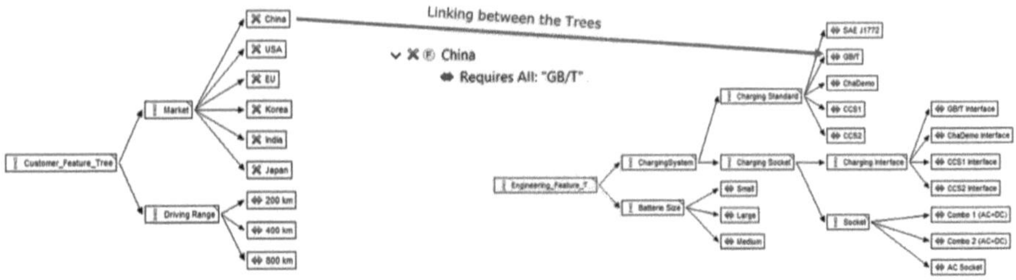

Fig. 19. Correlation between the Variability Description Models

With the use of the variability description, consistent variants can be created, because all the relations between the features are documented. This method can help to formalize expert knowledge in a model. This way, everyone can create a valid variant from the model. The combined sum of all variants can be seen as a solution space of all possible designs. We refer to it as a "150% model". By selecting the particular features in one variant, engineers can derive a particular 100% model again.

The variability description with the features in it exists additionally to the common MBSE artifacts from the RFLP concept. The traceability between the features and RFLP artifacts connection can be realized in different manners. Two possible approaches are presented and compared in the following (Wagner, et al. 2023).

- Direct Artifact Dependencies: This approach entails establishing a direct link from the feature to each corresponding element in all relevant RFLP viewpoints, as illustrated in Fig. 20.
- Indirect Artifact Dependencies: In this approach, the feature is exclusively connected to the requirements viewpoint, as shown in Fig. 21. The sequential variability-driven influence on elements within the functional, logical, and physical viewpoints occurs through the relationships denoted by << refine >> and << allocate >>.

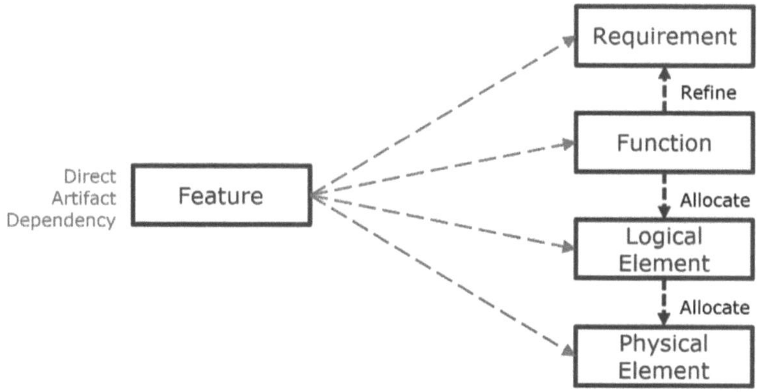

Fig. 20. Direct artifact dependencies, own representation according to (Wagner, et al. 2023)

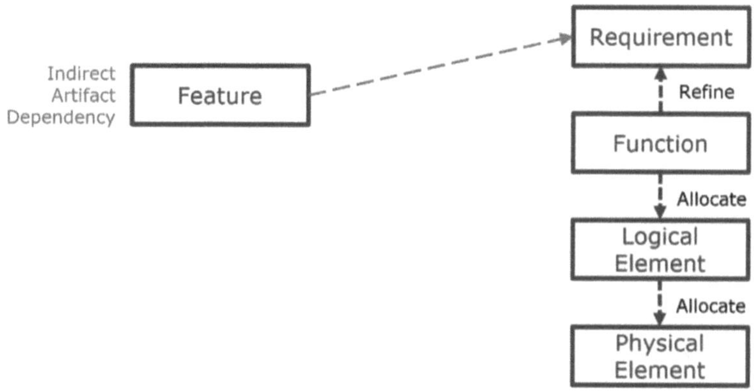

Fig. 21. Indirect artifact dependencies, own representation according to (Wagner, et al. 2023)

The method currently used at AVL is the direct artifact dependencies between the features and the RFLP artifacts. The variability description is linked to the relevant development asset in the V-model. To name some examples: Requirements documents, system architecture and specification in a SysML-tool, simulation libraries and test case databases.

The variability description is the central element that manages the 150% assets and filters for the needed features. Figure 22 shows the V-model and gives insight into the traceability concept. Especially related information on different system levels can be modeled in their interdependency.

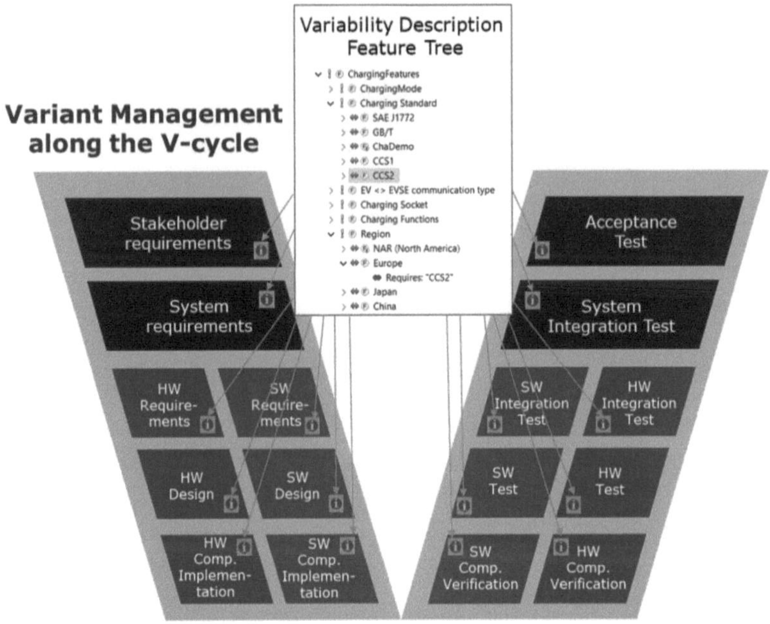

Fig. 22. Variability and different system levels in the V-model

The linking of information across system elements and system levels enables the use of the feature information in every deliverable. Based on the linking, new variants can be created together with their deliverables. In means of battery derivatives, this gives the opportunity to oversee the impact of a change of one feature.

Figure 23 shows a SysML internal block diagram of the thermal system of the battery. It includes not only one logical view but different solutions that can satisfy the function "cooling of the battery". The thermal conditioning can be done by a liquid cooling plate or liquid immersion cooling or by an air-cooled system. The liquid cooling plate and air-cooled system features are modeled in the diagram below. With the selection of one defined variant, one cooling method is selected and the not needed items in the diagram will be removed. The features are added as constraints to the system blocks and ports. The small blue V icons mark the items in the diagram that have a feature constraint.

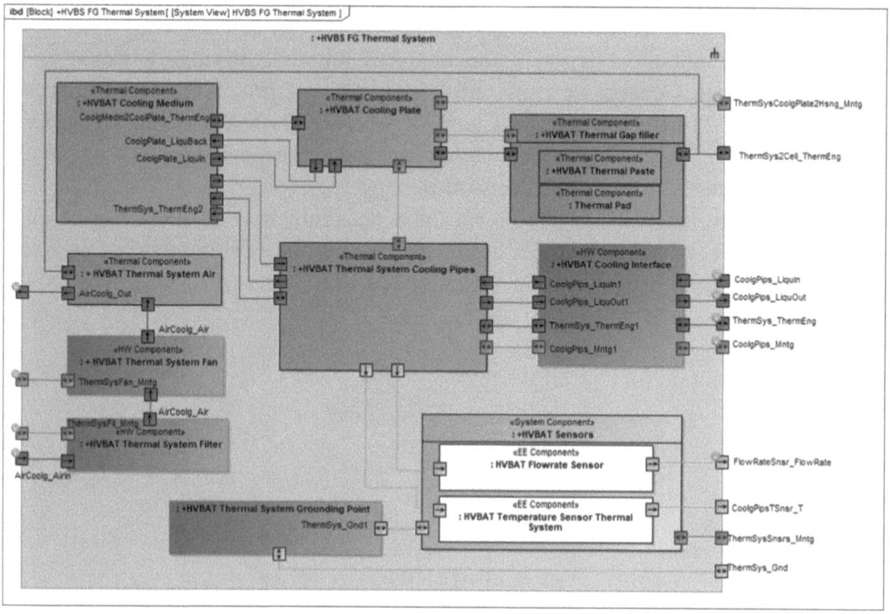

Fig. 23. Internal boundary diagram of the battery thermal system

To give an example, typically a lead variant and one derivative for a different market are developed. Then a third market creates a second derivative. Now the second derivative would be compared to the lead variant and the first derivate. The comparison is not limited to the variability description itself but also in the different development assets, which results in the comparison of requirements, architecture, and the corresponding tests. What changes if we take the new market regulations into account? What can be taken over from the existing development and what needs to be changed? The formalized variability description supports this analysis.

With the use of variant modeling, the reuse between the battery derivative projects can be improved and changes can be reused in all derivatives. This supports the exchange of lessons learned and best practices between the projects. However, manual documentation of information and their relationships is time-consuming and error prone. Hence, recent approaches of artificial intelligence (AI) have been investigated.

4.5 MBSE, KM and AI

As stated above, AI is an emerging technology which is considered to be crucial for advancements in technical systems and their development methodologies. In order to overcome the challenges of today's manual KM approaches, AI is being investigated as a potential solution. It is expected to assist the people that are involved in engineering processes in terms of

- knowledge reuse,
- project structuring and planning,
- storage and accessibility challenges.

"How can I create a chatbot for my business unit?" In contemporary times, this inquiry is arguably the most frequently encountered among knowledge workers.

In addressing this query, we explored the AI and enterprise knowledge approach illustrated in Fig. 24, which integrates ChatGPT with enterprise data using Azure OpenAI and Cognitive Search. These technologies are broadly expected to boost creativity, improve interactions, and facilitate decision-making.

This illustrative instance is contingent upon the utilization of the Azure OpenAI Service to gain access to the ChatGPT model (gpt-3.5-turbo), in conjunction with Azure Cognitive Search for the indexing and retrieval of data resources (Azure-Samples 2023).

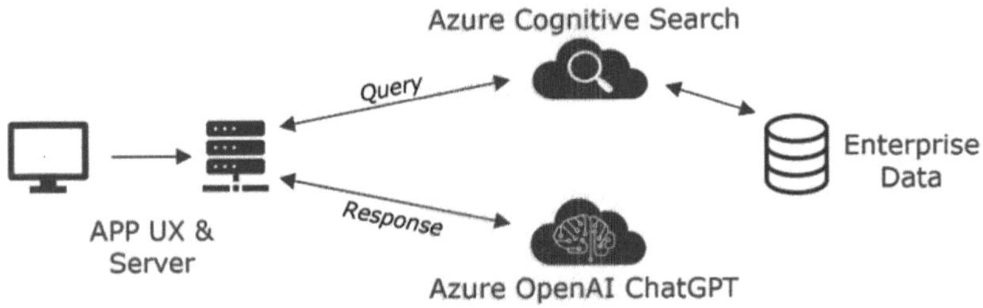

Fig. 24. ChatGPT + Enterprise data with Azure OpenAI and Cognitive Search Architecture, own representation according to (Azure-Samples 2023).

4.5.1 The Pattern

The architectural framework focuses on resolving enterprise search challenges by leveraging Microsoft Azure services, comprising two subsystems: the information querying subsystem and the data integration subsystem. This pattern allows interactive querying of structured and unstructured data using natural language, extracting relevant information in a presentation-friendly format. Key highlights include natural language queries, generating formatted responses enriched with references for transparency, usability enhancement through a ChatGPT API, and semantic search across structured and unstructured data. In the context of an engineering company, such data is cumulated in projects and comprises e.g. requirements, system specifications, test reports, among many others. The system considers intent, context, interrelationships, and conceptual associations, surpassing reliance on keyword alignment for more precise outcomes (Tanasseri 2023) (Fig. 25).

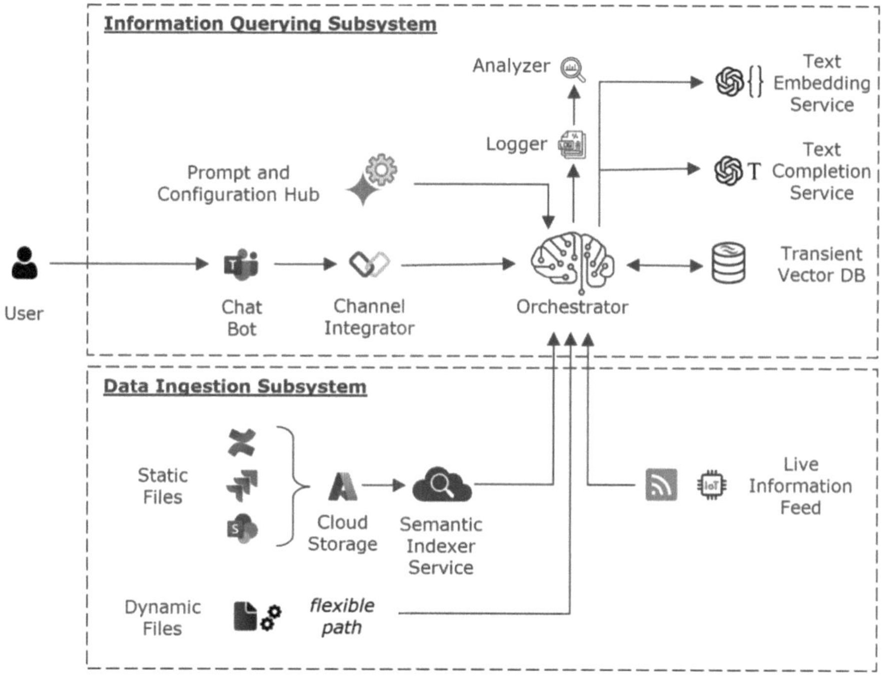

Fig. 25. Reference Architecture, own representation according to (Tanasseri 2023)

Following the AI based KMS scheme, the AVL IT team has spearheaded the effort to connect the internal enterprise data with Azure OpenAI and Cognitive Search, enabling employees to access relevant knowledge simply by posing questions to a ChatGPT-powered chatbot. The current iteration of the developed solution is based on Microsoft Copilot and is already available for all AVL employees. Figure 26 shows an exemplary user dialogue. The chatbot has access to all AVL documents which are in the user's cloud. In further work going on at the moment, this is intended to be extended to all documents while ensuring personal data rights as well as compliance with confidentiality agreements of the projects.

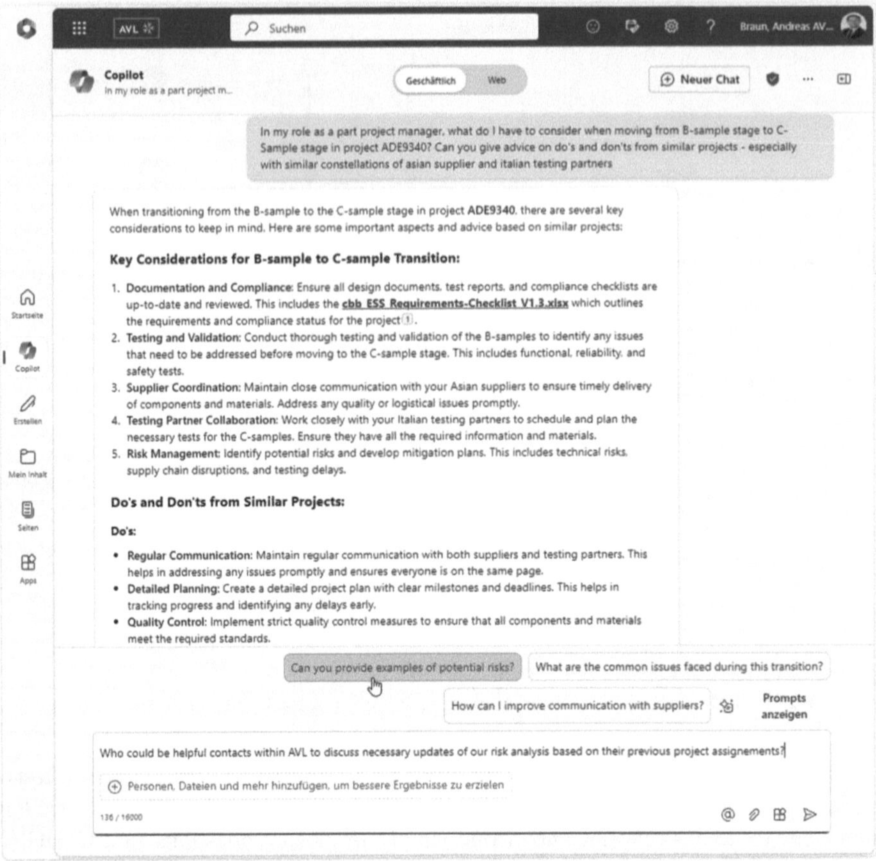

Fig. 26. exemplary user dialogue in the current implementation of AVL's Copilot

5 Conclusion and Outlook on Further Work

In conclusion, as we navigate the intricate journey from battery development as a black box to the establishment of a comprehensive enterprise socio-technical system, it becomes evident that addressing the identified challenges is crucial. (Braun, et al. 2021) Artifacts examined when studying a newly founded team (Braun & Löchl 2017) and based on their feedback (Novak, et al. 2021b) have proven to be well-suited to tackle the complexities that were identified in a pFMEA investigation. (Müller, et al. 2017).

Battery development encompasses a wide range of disciplines, from chemistry and thermodynamics to electronics, mechanical engineering, and software engineering. Managing organizational complexities, such as evolving requirements, varying standards, and global collaboration, is equally critical for success.

We have explored the intricate interplay of technological and organizational complexities within the realm of battery development, highlighting the crucial role of SE and MBSE in managing these challenges.

The application of MBSE at AVL has proven to be a valuable approach, offering a structured framework for managing requirements, functions, and logical aspects of the battery system. This approach not only aids in architectural design but also serves as a foundation for effective project communication and data management.

Furthermore, the incorporation of variant management within MBSE provides a means to address the increasing product variability in battery systems. This approach enables the systematic handling of diverse features and variants, ultimately leading to more efficient and knowledge-based development processes.

However, the journey doesn't end here. Notably, the existing limitations in usability within manual processes underscore the need for innovative solutions. (Badi, et al. 2024a, b) Recognizing the potential of AI technology, derived from the synthesis of data from findings (Badi, et al. 2024a, b) and the pertinent literature studies, constitutes a crucial perception.

The incorporation of AI into the development process presents a promising avenue for further improvement. AI has the potential to automate manual KM efforts, identify relevant development information, and ensure lessons learned are accessible and applicable across projects and teams. This smart technology can significantly enhance efficiency and knowledge transfer within the organization.

As we look ahead, it's clear that the integration of MBSE, KM, and AI holds the promise of streamlining battery development even further. By addressing the complexities at both the technical and organizational levels, enterprises like AVL can continue to innovate and deliver high-quality battery systems that meet the evolving needs of the market while maintaining a commitment to sustainability and efficiency.

In further work, AVL has set up a pilot team to elaborate the AI-supported battery engineering framework. Scientific studies carried out in parallel to the industrial projects in which the comprehensive approach is being applied will quantify the effectiveness of the AI-supported engineering framework.

Bibliography

Al-Kadi, O.S.: Knowledge management systems requirements specifications. University of Canberra, Faculty of Information Sciences and Engineering: s.n (2003)

Anderson, L.: Enterprise modeling for CubeSats. In: IEEE Aerospace Conference, vol. IEEE, pp. 1–16 (2014)

AVL. E-Mobility (2022). https://www.avl.com/en/e-mobility#:~:text=The%20nations%20of%20the%20world,the%20way%20to%20sustainable%20mobility. Accessed 07 Sept 2023

AVL, Success Story: AUDI Exceeds CO2 Fleet Targets (2022). https://www.avl.com/registration-cta?articleId=58441031&groupId=10138. Accessed 11 Sept 2023

Azure-Samples. Chatgpt + enterprise data with azure openai and cognitive search (2023). https://github.com/Azure-Samples/azure-search-openai-demo. Accessed 11 Sept 2023

Badi, I., Braun, A, Kallis, L., AI-supported model-based knowledge management in HV battery development. Münster, Haus der Technik (HDT) (2024a)

Badi, I., Braun, A., Kallis, L.: Model-based knowledge management in HV battery development. Curran Associates Inc, Straßburg (2024b)

Bellamine, N., Ghézala, H.B.: On the convergence of collaboration and knowledge management. arXiv preprint arXiv:1202.6104 (2012)

Braun, A., Löchl, E.: Ein Lean Management Ansatz für die Entwicklung von Traktionsbatterien. In: Maurer, M., Schulze, S. (eds.) Tag des Systems Engineering 2017, pp. 267–270. Carl Hanser Verlag, Paderborn (2017)

Braun, A., Maletz, M.: Improved Battery Performance by Model-Based Systems Engineering. Aachen, Haus der TEchnik (HDT) (2015)

Braun, A., Schiffbänker, P., Falls, B., Küpper, K.: Case study: battery system development. Systems Engineering for Automotive Powertrain Development, pp. 703–726. Springer, Cham (2021)

Braun, A., Schiffbänker, P., Falls, B., Küpper, K.: Case study: battery system development. In: Hick, H., Küpper, K., Sorger, H. (eds.) Systems Engineering for Automotive Powertrain Development. Powertrain. Springer, Cham (2021). https://doi.org/10.1007/978-3-319-99629-5_20

CSE Aachen GmbH, 2020. *KBM* - Die Erwartungen wurden übertroffen. https://cse.rwth-campus.com/unkategorisiert/die-erwartungen-wurden-uebertroffen/. Accessed 27 Nov 2023

Denger, A.: Organisationaler Wandel durch die Emergenz Cyber-Physikalischer Systeme: Die Fallstudie AVL List GmbH. HMD Praxis der Wirtschaftsinformatik **51**, 827–837 (2014)

Einkopf, J., Braun, A.: Minimizing risks in battery design. Curran Associates Inc, Mainz (2017)

Hennige, V., Braun, A.: Complexity of batteries for electric vehicles – and how to manage it!. Dresden, Fraunhofer-Institut für Keramische Technologien und Systeme IKTS (2015)

INCOSE, What is Systems Engineering? (2019). https://www.incose.org/systems-engineering. Accessed 27 Nov 2023

Kleiner, S., Kramer, C.: Model based design with systems engineering based on RFLP using V6. In: Abramovici, M.S.R. (ed.) Smart Product Engineering: Proceedings of the 23rd CIRP Design Conference, Bochum, Germany, March 11th-13th, 2013, pp. 93–102. Springer, Berlin Heidelberg (2013)

Lindemann, U., Maurer, M., Braun, T.: Structural complexity management: an approach for the field of product design. Springer, Berlin, Heidelberg (2009)

Maletz, M.: SEPA Method Description for Systems Engineering, s.l.: AVL (2021)

Maletz, M., Bajzek, M., Hick, H.: Systems Engineering Methods for Automotive Powertrain Development. Systems Engineering for Automotive Powertrain Development, pp. 403–442. Springer, Cham (2021)

Minke, R.: Stakeholder perspectives for Model-Based-Systems-Engineering in battery development. Graz: AVL List GmbH (available from AVL upon request) (2015)

Müller, M., Schiffbänker, P.A.A., Braun, A.B.N.: Efficient application of MBSE using reference models: a PGE case study. Curran Associates Inc, Vancouver (2017)

North, K.: Wissensorientierte Unternehmensführung. Springer Gabler, Wiesbaden (2016)

Novak, et al.: Battery derivative development using model based systems engineering. Wiener Motoren Symposium 2021 (2021a)

Novak, W., et al.: Battery derivative development using model based systems engineering. Austrian Society of Automotive Engineers (ÖVK), Wien (2021b)

Nunamaker, J.F., Romano, N.C., Briggs, R.O.: A framework for collaboration and knowledge management. In: 34th Hawaii International Conference on System Sciences (2001)

Qureshi, S., Hlupic, V., Briggs, R.: On the convergence of knowledge management and groupware. CRIWG'04, pp. 25–33 (2004)

Sandhu, S., Tripathy, B.: Knowledge management using recommender systems. In: Anouncia, S.M., Wiil, U.K. (eds.) Knowledge Computing and its Applications, 1st edn., pp. 183–207. Springer, Singapore (2018)

Simon, H.A.: The architecture of complexity. The Sciences of the Artificial, pp. 192–229 (1969)

Tanasseri, N.: Open AI RAG (Retrieval Augmented Generation)- Architecture Pattern for Enterprise Information Management (2023). https://www.linkedin.com/pulse/open-ai-architecture-pattern-enterprise-information-namit-tanasseri/. Accessed 07 Sept 2023

Wagner, H., Rössler, W., Zuccaro, C.: How to structure variability in large-scale complex engineered systems. Tag des Systems Engineering 2023 (2023)
Yang, L., Cormican, K., Yu, M.: Ontology learning for systems engineering. IEEE Trans. Industr. Inf. **17**, 1039–1047 (2020)

Methodological Implications of Co-evolution in the Generative Design Process

Vishal Singh(✉) and Divyanshu Bajpai

Indian Institute of Science, Bangalore, India
singhv@iisc.ac.in

Abstract. This chapter focuses on the co-evolutionary aspects of the generative design process, highlighting the need for specific research into the interdependence between the knowledge types, thinking paradigms, and the design systems under consideration. The chapter argues that, in addition to the co-evolutionary aspects of a design process in general, there are unique co-evolutionary aspects associated with the generative design process. These additional co-evolutionary aspects unique to the generative design process arise from the need to represent the primary design problem in a generative design form and to design and implement a generative design system to support solving the primary design problem. The chapter is based on theoretical arguments, reinforced with evidence from a preliminary set of design experiments conducted using a tangible design kit.

1 Introduction

A comprehensive and expansive exploration of design space is one of the fundamental expectations in a creative design process. In a creative design process, the search through the design space continues until the 'satisficing' (Simon 1969) conditions are met. This exploration is known to be knowledge-intensive and iterative. The exploration and exploitation, divergence and convergence, synthesis and analysis, formulation and reformulation, and the overall understanding of the problem and the solution occur back and forth through a co-evolutionary process (Schön 1992, Poon and Maher 1997, Dorst and Cross 2001). Consequently, design processes can be highly demanding. Accordingly, various design methods and tools have been developed to aid design processes. Generative design (GD) system (Krish 2011, Singh and Gu 2012) are one such system which supports design exploration and design automation.

Several authors (e.g. Kari 2005, Stiny 1980) have noted that a substantial amount of knowledge is required to describe the GD process and implement a GD system. This observation echoes researchers such as Frankenberger and Badke-Schaub (1999) and Rasmussen (1985), who emphasise information availability as a central factor for successful design outcomes. Drawing upon Venselaar K et al. (1987), Smith and Duffy (2001) presented a design reuse model consisting of processes for design by reuse, domain exploration, design for reuse, and six knowledge-related components. Smith and Duffy (2001) also developed a design knowledge taxonomy.

Therefore, recognising the reliance of GD on the utilisation of multiple knowledge types and following a review of the literature, it is argued that at least three types of

K. Gericke et al. (Eds.): DSC 2022/2023, PTT, pp. 182–196, 2026.
https://doi.org/10.1007/978-3-032-16278-6_11

knowledge are required for GD implementation, including domain knowledge, representational knowledge, and geometrical knowledge. Domain knowledge is related to the problem context. For example, architecture or mechanical. Representational knowledge is related to problem formulation, abstraction and representation. For example, in mathematics, the same problem can be formulated, abstracted and represented in different ways, such as a differential or algebraic form, requiring corresponding knowledge about the representational form–calculus or algebra. In GD, any given problem must be formulated, abstracted and represented in one form or another, requiring knowledge of GD representations. Geometrical knowledge refers to the mathematical and formal knowledge associated with geometrical representation, operations and manipulations needed to arrive at the formal and embodiment aspects of a design, typically done in a one-, two- or three-dimensional space. For example, geometrical knowledge would include reasoning about shapes, coordinates, sizes, etc. and their relationships. In engineering design, including Computer-Aided Design (CAD), geometrical representation and reasoning are typically considered an essential part of design process, even if a significant amount of reasoning about the geometry may occur in the symbolic reasoning space.

It is argued that the GD process evolves as a negotiation between the domain knowledge, the representational knowledge, and the geometrical knowledge. This negotiation is mediated by a combination of design thinking (Brown 2008, Cross 2011, Dorst 2011), systems thinking (Arnold and Wade 2015, Forrester 1987, Meadows 2015), and computational thinking (Wing 2006, 2008). As noted by several authors, design thinking, systems thinking, and computational thinking are intertwined (Singh 2025, Shute et al. 2017, Kelly and Gero 2021). This intertwining should be explicitly observable in a GD process because of the inherent requirements of explicit knowledge representation in the GD system. Hence, besides the usual coevolution of problem and solution that are observed in any creative design process, there are at least two notable kinds of coevolution that are expected in a GD process as problem and solution gain specificity: (1) articulation and application of the domain knowledge, representational knowledge, and geometrical knowledge relevant to the problem at hand, and (2) articulation and application of design thinking, systems thinking, and computational thinking aspects of the problem at hand.

Accordingly, the rest of the chapter is organised in the following order. The initial sections are based on theoretical arguments, which are later tested and reinforced by evidence from a preliminary empirical study. First, a discussion on the three types of knowledge is presented, including their interdependencies and articulation and application during the GD process. Second, the three thinking paradigms are discussed, and how they intertwine and coevolve during the GD process is reviewed. Third, the connection between the knowledge types and the thinking paradigms is discussed. After that, preliminary empirical research and its findings are described, showing how the empirical evidence supports the claims of the discussed coevolutionary aspects of the GD process. The chapter concludes with a brief discussion of the methodological implications of intertwining knowledge types and thinking paradigms reviewed in this chapter.

An experiment was conducted to test and validate the utilisation of these knowledge types in GD. The observations from the experiments confirmed the explicit and implicit use of the three types of fundamental knowledge: domain knowledge, representational knowledge, and geometrical knowledge. A process of co-evolution is also observed during the experiments.

2 Explicit Knowledge Types in the GD Process

A GD implementation project starts with an intended purpose, typically associated with a specific domain. Therefore, the initial steps in planning the GD implementation involve understanding the problem context, which requires domain knowledge. The domain knowledge allows the GD system implementer to articulate the design objectives and the domain-relevant parameters, variables, relationships and dependencies, conditions, constraints, assumptions, and rules. This articulation of the required knowledge rarely happens in one go. As the implementation progresses, the GD system implementer typically refines the model through iterative steps. In between these iterations about the domain knowledge, the implementation requires formulating and representing the domain knowledge in a computational form, which requires knowledge of the different representational forms, the corresponding modelling techniques, and the strengths and weaknesses of the alternative representational methods. The representational form includes the conceptual form, for example, a cellular automaton (CA), shape grammar (SG) or node-link, and the corresponding computational representation, which translates the conceptual form into a parametric and computable form, typically with software codes, syntax, structure, and integration with relevant algorithmic approaches. In addition, in engineering and architectural design, the output of the GD systems is often expected in terms of the CAD models, which require knowledge of geometry and geometrical modelling. Like the necessary iterations in articulating domain knowledge from early conceptualisation to detail design, iterations are required in the articulation of representational knowledge as well as geometrical knowledge. The intertwined iterations between domain knowledge, representational knowledge and geometrical knowledge are represented in Fig. 1.

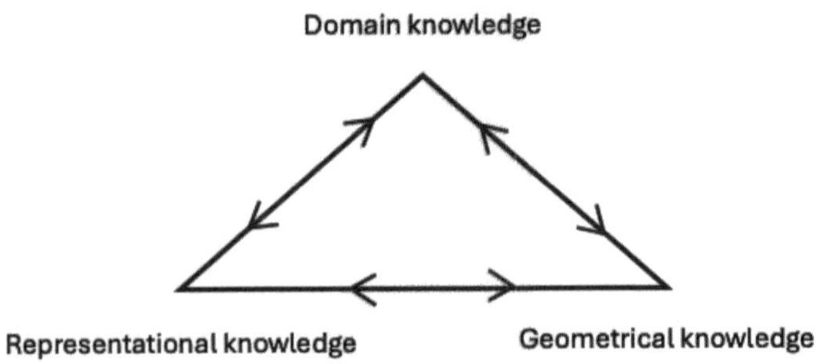

Fig. 1. The iterative flow between domain, representational and geometrical knowledge in the GD process (arrows signify mutual dependence and co-evolution)

3 Thinking Paradigms and the GD Process

As noted earlier, the intertwining of design thinking, systems thinking, and computational thinking should be more explicitly observable in the GD process, given the need for explicit knowledge capture and representation in a GD implementation. At the same time, implementing a GD system (GDS) involves multiple levels of design reasoning involving multiple design systems (see Fig. 2). The primary design system (PDS) is concerned with the problem domain for which the design solution is sought. For example, the primary problem could be designing an urban system for which a GD system is to be implemented. The secondary design system is concerned with the GD system, which must meet its purpose of enabling the design exploration for the primary (e.g. urban) design system. For both design systems, meta-level reasoning is required, which determines the approach and strategy towards problem formulation and solution generation for each design system. While the meta-level reasoning for formulating and solving the primary design system (e.g. urban system) is partly captured in the design of the secondary design system (the GD system), the meta-level reasoning in the design of the GD system may remain implicit with the GD system implementer unless documented explicitly. For example, a well-commented software code may explicitly capture some meta-level reasoning about the design of the GD system.

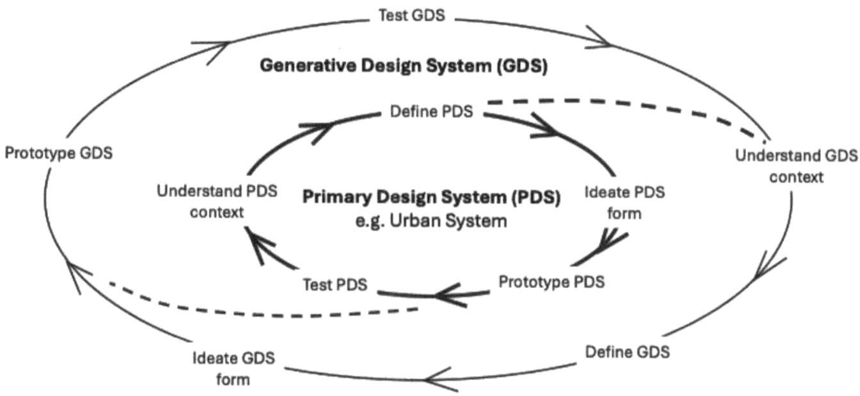

----- Dotted lines indicate interdependencies between PDS and GDS

Fig. 2. Iterative design thinking with a co-evolving understanding of problem and solution for the primary design system and the generative design system

Figure 2 schematically shows how the design of the GD system for solving a domain-specific problem using that GD system follows the design thinking process, where design thinking is required for both the PDS and the GDS as a co-evolutionary process.

Further, as noted by default in the nomenclature of PDS and GDS, the GD system implementer explicitly reasons about the systems, applying systems thinking implicitly or explicitly. For the PDS, the GD system implementer must develop an understanding of the problem context, including an assessment of aspects such as the: (1) System components and configuration – what constitutes the problem context, both internal and external to the system under consideration, (2) System boundary – what should be considered and what can be left out, and (3) System dynamics – what are the interdependencies and interrelationships. Similarly, for the GDS, the implementer must assess the required system components and configurations – variables, parameters and other elements in the GD model, system boundary – scope of the GD system, and the system dynamics – rules, functions and interdependencies in the GD model. The systems thinking for the PDS and GDS also co-evolves iteratively as the GD implementer iterates through the different stages of the development and refinement of the GD system (see Fig. 3).

The relevance of computational thinking to implementing a GD system is well documented (Menges and Ahlquist 2011). There are four broad aspects of computational thinking, including: (1) decomposition and re-composition, (2) abstraction and representation, (3) algorithms and steps, and (4) pattern recognition. Each aspect is explicitly noticeable in the iterative implementation of a GD system. For instance, a GD system would include: (1) a decomposition of the system to be implemented in terms of the model elements, (2) abstraction and representation in the form of one or the other GD representation, which includes only the most relevant details through abstraction, (3) algorithms and steps involved in running the GD system, and (4) identification of patterns leading to the GD system development, and seeking to identify patterns that emerge from running the GD system. Similarly, the computational thinking aspects are equally applicable to the iterative co-evolution in the understanding of the problem and solution associated with the PDS. For instance, identifying the elements in the problem context and the interdependencies between them requires decomposition and pattern recognition. Defining the purpose and scope of the design involves abstraction and representation. Understanding the problem context, the patterns, and the decomposition requires iterative steps. Thus, the computational thinking considerations for the GDS and the PDS must also co-evolve through iterations and refinements (see Fig. 4).

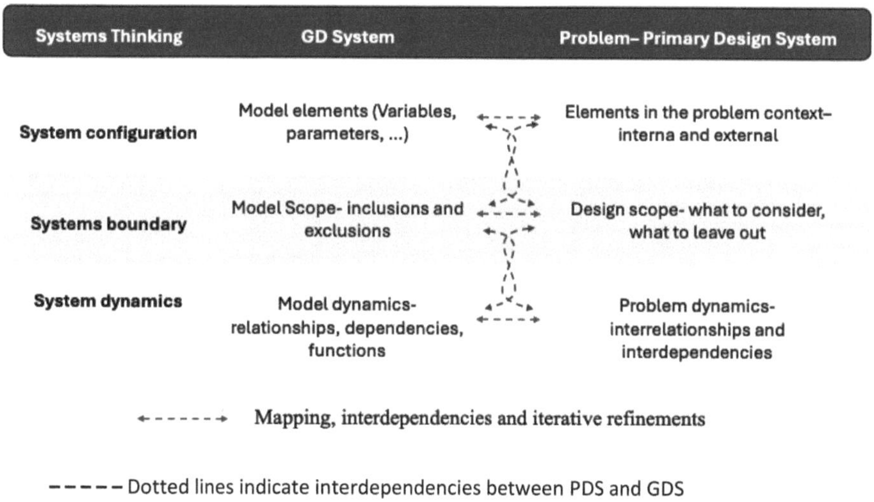

Fig. 3. Iterative Systems thinking with a co-evolving understanding of the GDS and PDS

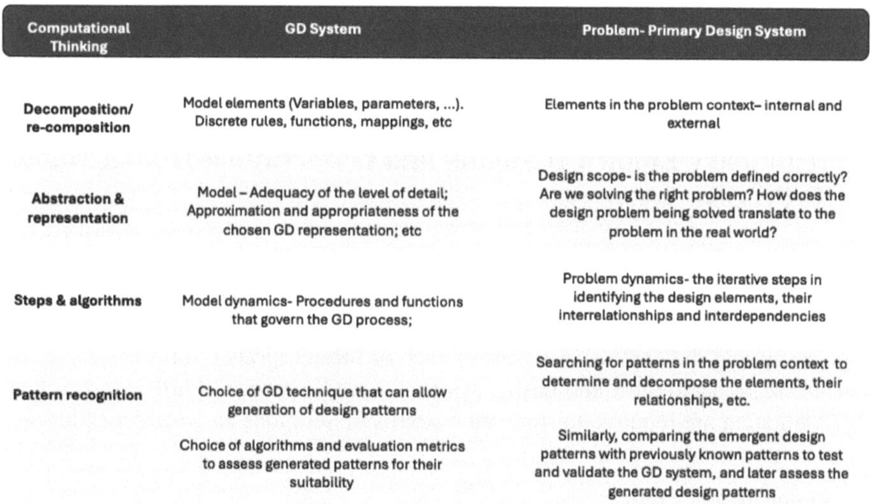

Fig. 4. Iterative Computational thinking with a co-evolving understanding of the GDS and PDS

4 The Connection Between the Knowledge Types and Thinking Paradigms in the GD Process

So far, we have separately discussed the iterative flow between the different knowledge types, between the different thinking paradigms, and between PDS and GDS. However, these iterative cycles within the knowledge types, thinking paradigms and design systems cannot be mutually exclusive. Consequently, there would be complex dependencies across aspects of a GD process, which is inadequately understood and researched so far.

Figure 5 provides one such framework that can be used to design experiments and studies to investigate the interdependencies between the knowledge types, thinking paradigms and design systems in a GD process.

A pairwise mapping and review for each aspect is desirable. The specific decomposition can vary. For instance, instead of using the five steps from the d-school Design Thinking framework, the FBS ontology from Gero's (1990) design prototype could be used.		Design Thinking					Computational Thinking				Systems Thinking		
		Understand	Define	Ideate	Prototype	Test	De/Re-composition	Abstraction/ Representation	Steps & Algorithms	Pattern recognition	System elements and configuration	System boundary	System Dynamics
Domain knowledge	Primary Design System (PDS)												
	GD System (GDS)												
Representational knowledge	PDS												
	GDS												
Geometrical knowledge	PDS												
	GDS												

Fig. 5. A potential framework for investigating the interdependencies between knowledge types, thinking paradigms, and design systems under consideration in a GD process

5 Preliminary Empirical Studies and Observations from a Manual GD Process

A preliminary set of basic experiments was conducted to observe the methodological aspects of the GD process, especially the iterations between the knowledge types, thinking paradigms, and design systems. The scope of these experiments was limited to observing the explicit instances of each of these aspects – different knowledge types, thinking paradigms, and design systems in a GD process. Further research and experimentation are required to study the patterns in iterations and interdependencies.

5.1 Experiment Details

Design students are introduced to two GD techniques using a hands-on (Cassin 2013, Junginer 2007) and a tangible toolkit (Marshall 2007) for a design problem. The two techniques are Shape Grammar (SG) (Stiny 1980) and Cellular Automata (CA) (Wolfram 1983, Kari 2005, Herr and Ford 2016). The students generate the required design using the two different techniques in two separate sessions. A think-aloud protocol (Fonteyn et al 1993) is used in the design sessions, and the data is video recorded on tape, capturing both the verbal data and the evolution of design using the tangible kit. The study aimed to identify the patterns in the students' design thinking process while applying the GD techniques to solve a given problem.

An exhibition layout design problem is used for design experiments. The experiments were conducted with six design students at the postgraduate level. Four of these students have a background in product design, and two in architecture. None of the subjects have prior experience with GD. Therefore, for each module involving a GD process, a demonstration video and reference manual explaining how to perform activities are provided before the experiment sessions. The researchers clarify all the doubts about the techniques before the experiments. The researchers observing the design sessions are also available to clarify doubts during the experiment sessions.

The stated problem is: "*To design an exhibition space layout to display paintings such that there are four exhibition zones with similar display capacity. The paintings are displayed on stands such that there are adequate numbers of surfaces to display the paintings. Adequate space must be provided inside each zone for visitors to walk and view the paintings. The layout will be done in a 2D space (top view)*".

5.2 Requirements/Constraints on the Display Stand

A display stand combines flat panels as surfaces for displaying the paintings. The top view of a standard display panel can be abstracted and represented as a linear element. Consequently, in these experiments, a display panel is represented by a tangible linear object, a matchstick. A display stand is a set of multiple connected matchsticks. The viewing angle should be between 90 and 180 degrees. These panels are to be placed so that both the sides of the panel – front and back – can be used for displays.

Three types of experiments are conducted using the same building blocks – matchstick. Throughout the experiment sessions, the designer is asked to think aloud. A video camera pointed towards the tangible design surface captures all the design steps and the designer's actions and moves. Thus, the decision-making process and the design actions throughout the design sessions are recorded using audio-visual media.

5.2.1 Experiment Scenario 1: Freeform Design (FT) Using Tangible Objects Without Predefined Rules

There are no predefined rules, additional constraints, or limitations on the design generation process. Participants can create any configuration of the display structure and exhibition layout using the panels (matchsticks) if it meets the constraints defined in the problem statement. Clay is provided as joining material. Figure 6 shows an example layout generated by one of the participants.

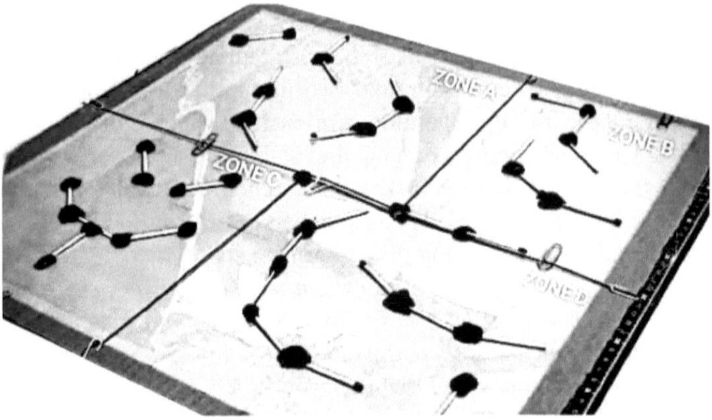

Fig. 6. Exhibition layout designed without any pre-defined design rules

5.2.2 Experiment Scenario 2: Design Using Cellular Automata and Tangible Objects (CAT)

The constraints in these experiments include: (1) the design must use CA principles and (2) the maximum number of panels in a single display structure cannot be more than five. Since CA is usually performed on a grid pattern, a grid is provided to participants, along with the building block (matchsticks), cell states, and basic CA rules that could be used to generate the design manually, Figure 7 The design activity is performed in two stages of system decomposition, starting with a bottom-up process. In the first stage, the design of a display structure is finalised. In the second stage, the overall layout plan with different zones is finalised.

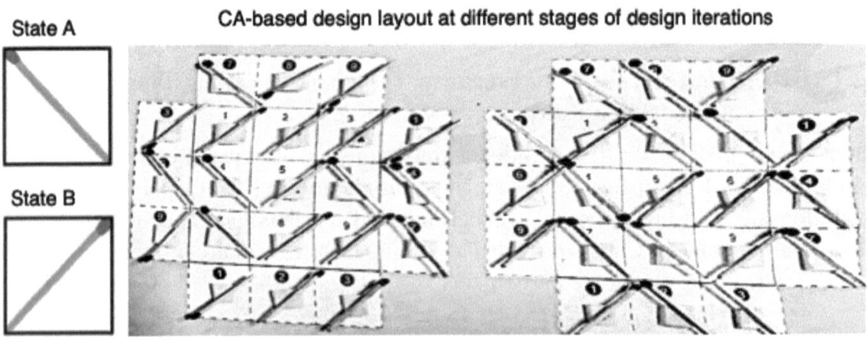

Fig. 7. Possible cell states and emerging design outcomes for one of the participants

5.2.3 Experiment Scenario 3: Design Using Shape Grammar and Tangible Objects (SGT)

The design problem remains the same, but shape grammar is the chosen generative technique. Shape grammar starts with an initial shape (matchstick) and predefined rules. For example, a sample generation rule is shown in Fig. 8.

Two constraints are to be followed in the design of the display structures: (1) the angle between any two panels should be between 90 and 180 degrees, and (2) the number of panels in a display structure should be less than or equal to five. Once again, the design activity is performed in two stages of system decomposition, following a bottom-up process – first, the design of display structures and then the overall layout plan. Some of the generated display structures derived using shape grammar are shown in Fig. 8.

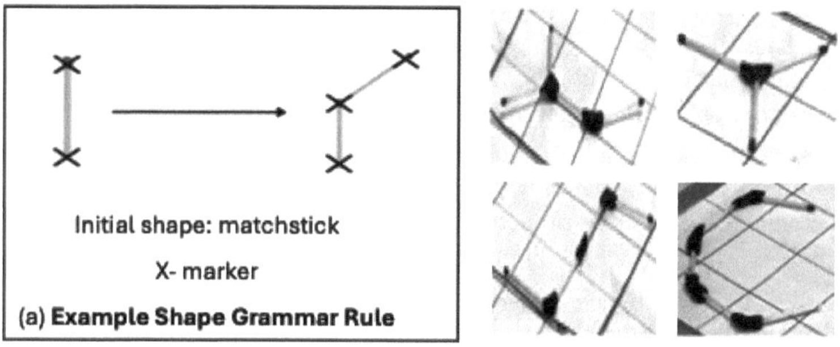

Fig. 8. Sample shape grammar rule and generated display structure alternatives

6 Post-experiment Interview

In addition to the data collected from the think-aloud protocol and video recordings of the design progression, 10-min post-experiment semi-structured interviews were conducted with the participants. The interview questions are outlined in Table 1.

Table 1. Questions for semi-structured interviews

Interview Questionnaire	Why was this question asked?
Which techniques did you find easier to use? Why?	To assess (1) abstraction and representation (2) mapping between PDS and GDS (3) ease of thinking in terms of rules, procedures and algorithmic steps
Any tangible object(s) you felt missing from the experiment kit that would have helped you during the experiment?	To check if the participants have anything to add about PDS and GDS in terms of (1) decomposition (2) systems thinking (3) pattern recognition
Which technique granted you more freedom in designing? Rank the three sub-experiments in terms of the freedom to design	To assess (1) abstraction and representation (2) mapping between PDS and GDS (3) ease of thinking in terms of rules, procedures and algorithmic steps
Which aspects of the experiment were easy to understand, and which parts were difficult?	To qualitatively understand the complexity and cognitive load in managing interdependent aspects in a GD process

7 Observations, Results and Findings from the Hands-On Experiments

Since the participants were given matchsticks as tangible objects to design the layouts, the research team had already done part of the PDS problem formulation, decomposition, abstraction, and representation. Hence, the participants did not have to start from scratch about designing the layout. The given elements also defined the scope for design exploration, limiting the options of the display blocks to be linear elements of a given size. Thus, from the outset, some aspects of design thinking, systems thinking, and computational thinking are outlined both at the PDS and GDS levels.

In experiment 1, even though the elements are given, no rules or dependencies are predefined other than the design constraints. Participants could generate free-form designs or formulate design generation rules or procedures. None of the six participants developed any design generation rules or procedures, which meant that the solutions generated in experiment 1 did not follow a GD process despite the repetition of the same building block. While it was expected that the participants would feel least constrained and be most creative in such a free-form design session, almost all participants reported being a little lost and directionless in experiment 1 compared to the clarity of purpose in experiment sessions 2 and 3. From the perspective of the design process, the iterations were mainly about the domain knowledge (about exhibition design and displays), with occasional references to the geometrical knowledge (about viewing angles, distances, etc.). The explicit reference to representational knowledge was limited and ad-hoc, if at all. Overall, the participants reportedly found the freeform design session more challenging, open-ended, and ambiguous even though they recognised that it offered the most opportunities to be creative. The participants noted that the GD approach in experiments 2 and 3 structured their design process, allowed them to decompose the problem in terms of the system and sub-systems, and gave the design process a more definitive direction.

In experiment 2, the participants demonstrated the ability to apply CA rules to generate solutions for display structures and the overall exhibition layout. The discussions and deliberations around the dependence between the orientation of the display structures, the flow of people, and visibility angles demonstrated the interdependence and iterative consideration for domain knowledge, representational knowledge and geometric knowledge. However, most participants also found it challenging to keep track of the computational requirements for checking the validity of all the cell states at each iteration cycle. Thus, when using CA, they found it challenging to maintain coherence between PDS and GDS.

In experiment 3, the participants demonstrated the most comfort and ability to apply SG rules to generate solutions for display structures and the overall exhibition layout. The discussions and deliberations around the dependence between the orientation of the display structures, shape manipulations, the flow of people, and visibility angles demonstrated the interdependence and iterative consideration for domain knowledge, representational knowledge and geometric knowledge. Unlike the challenges faced in the CA usage, the participants felt comfortable using SG. According to the participants, the application of SG rules seemed most natural to the PDS because creating display structures using standard elements is nothing but shape manipulation without the additional zoning constraints of a cell and grid, which was posed by CA. This discussion once again reflects the interdependence of the PDS and GDS. When using SG, the participants found it easy to maintain coherence between PDS and GDS.

In addition to the observations and direct feedback from the participants, the collected audio-visual data was transcribed for thematic analysis (Braun and Clarke 2006). Thematic analysis is an effective tool for identifying, analysing, and reporting patterns or themes within the data. The results and insights from thematic analysis complement the findings from the post-experiment interviews.

Four significant discussion themes associated with the design experiments emerged from the thematic analysis, including: (1) Participants' attention to user/visitor experience, (2) Participants' perception and comprehension of the problem, (3) Participants' views on GD, and (4) Participants' view on the role of tangible GD experiment kit. The themes are briefly described in Table 2.

Table 2. Themes and associated codes

Theme	Description	Codes under each theme
Attention to User/ Visitor Experience	Four out of six participants put considerable focus on the walking space arrangement of the panel, the field of view, and the angle between panels to improve the visitor experience	Space for walking (Geometry, Domain) Panel arrangement (Geometry, Domain) Gallery theme experience (Domain) Field of view (Geometry, Domain) Angle between panels (Geometry)
Perception and comprehension of the problem	All participants also spent considerable time explaining and seeking clarifications on how they interpreted and understood the problem in their own way	Exploration of design space (Domain, representation) Reinforcing problem objectives (Domain) Doubt clarification (Domain, representation) Constraint understanding (Domain)
Views on GD	Participants expressed their views on the specific GD techniques, including their intuitiveness and ease/ difficulty in application	Understanding rules (Domain, representation) Intuitiveness (Domain, representation) Design Freedom (Domain, representation) Computational thinking (Domain, representation)

(*continued*)

Table 2. (*continued*)

Theme	Description	Codes under each theme
Views on the role of tangible GD kit	Participants shared the experience of learning and applying GD with the help of a tangible kit	Geometry knowledge (Geometry) Panel size (Geometry, Domain) Space and real estate (Geometry, Domain) Visualizing spatial capacity (Geometry, representation)

The respondents report a positive experience learning and applying GD using a tangible kit. Participants reported and demonstrated satisfactory acquisition and comprehension of GD concepts, unconstrained by the fear of a lack of scripting or programming knowledge. The feedback suggests that the use of a tangible kit allowed adequate design thinking, systems thinking, and computational thinking for the PDS while reducing the level of design thinking, systems thinking, and computational know-how that is required for developing and implementing a GDS on the digital platform. The participants need not worry about what constitutes a working software program, how to create and execute the software system architecture, prototype and refine it, and so on. This freedom allowed them to focus more on conceptual mapping between the PDS and GDS.

In summary, the preliminary experiments reveal interdependencies and iterations between the knowledge types, thinking paradigms, and design systems in a GD process. Given the diverse types of knowledge, multiple aspects and frameworks for each thinking paradigm, and multiple levels of design systems to consider, the interdependencies create a complex iterative flow in the GD process. This complexity is exhibited in Fig. 5. Thus, from a methodological perspective, there is a need to investigate the GD process in greater detail to identify and map the complex iterative flow between the different aspects and identify patterns that can be generalised and simplified into a coherent model of coevolution of knowledge types, thinking paradigms, and design systems in a GD process.

8 Conclusion

This chapter emphasises the coevolution of knowledge types, thinking paradigms, and design systems in a GD process. Based on theoretical arguments supported by preliminary empirical findings, the chapter highlights the methodological implications and need for research to investigate these complex iterative flows and co-evolutionary aspects in a GD process, especially connecting the primary design system and the GD system.

References

Arnold, R.D., Wade, J.P.: A definition of systems thinking: a systems approach. Procedia Comput. Sci. **4**, 669–678 (2015)

Braun, V., Clarke, V.: Using thematic analysis in psychology. Qual. Res. Psychol. **3**(2), 77–101 (2006)

Brown, T.: Design thinking. Harv. Bus. Rev. **86**(6), 84–92 (2008)

Cassim, F.: Hands-on, hearts-on, minds-on: design thinking within an education context. Int. J. Art Des. Educ. **32**(2), 190–202 (2013)

Cross, N.: Design thinking: understanding how designers think and work. Berg (2011)

Dorst, K., Cross, N.: Creativity in design process: co-evolution of problem-solution. Des. Stud. **22**, 425–437 (2001)

Dorst, K.: The core of 'design thinking' and its application. Des. Stud. **32**(6), 521–532 (2011)

Fonteyn, M.E., Kuipers, B., Grobe, S.J.: A description of the think-aloud method and protocol analysis. Qual. Health Res. **3**(4), 430–441 (1993)

Forrester, J.W.: Lessons from system dynamics modelling. Syst. Dyn. Rev. **3**(2), 136–149 (1987)

Frankenberger, E., Badke-Schaub, P.: Information management in engineering design–empirical results from investigations in industry. In: Proceedings of ICED, pp. 911–916 (1999)

Herr, C.M., Ford, R.C.: Cellular automata in architectural design: From generic systems to specific design tools, Automation in Construction, vol. 72, pp. 39–45. Elsevier BV (2016)

Junginger, S.: Learning to design: giving purpose to heart, hand and mind. J. Bus. Strateg. **28**(4), 59–65 (2007)

Kari, J.: Theory of cellular automata: a survey. Theoret. Comput. Sci. **334**(1–3), 3–33 (2005)

Kelly, N., Gero, J.: Design thinking and computational thinking: a dual process model for addressing design problems. Des. Sci. **7**, E8 (2021). https://doi.org/10.1017/dsj.2021.7

Krish, S.: A practical generative design method. Comput. Aided Des. **43**(1), 88–100 (2011)

Marshall, P.: Do tangible interfaces enhance learning? In: Proceedings of the 1st International Conference on Tangible and Embedded Interaction, pp. 163–170 (2007)

Meadows, D.H.: Thinking in systems. Chelsea Green Publishing (2015)

Menges, A., Ahlquist, S.: Computational design thinking: computation design thinking. Wiley (2011)

Poon, J., Maher, M.L.: Co-evolution and emergence in design. Artif. Intell. Eng. **11**(3), 319–327 (1997)

Rasmussen, J.: The role of hierarchical knowledge representation in decision making and system management. IEEE Trans. Syst. Man Cybern. **SMC-15**(2), 234–243 (1985). https://doi.org/10.1109/TSMC.1985.6313353

Schon, D.: Designing as a reflective conversation with the materials of a design situation. Knowl.-Based Syst. **5**, 3–14 (1992)

Shute, V.J., Sun, C., Asbell-Clarke, J.: Demystifying computational thinking. Educ. Res. Rev. **22**, 142–158 (2017)

Singh, V., Gu, N.: Towards an integrated generative design framework. Des. Stud. **33**(2), 185–207 (2012)

Singh, V.: The interaction between design thinking, systems thinking, and computational thinking. In: Chakrabarti, S., Acharya (eds.) The Future of Design Education: Proceedings of InFuSED23 (2025)

Smith, J.S., Duffy, A.H.B.: Re-using knowledge: why, what and where. In: 13th International Conference on Engineering Design (ICED 01) (2001)

Stiny, G.: Introduction to shape and shape grammars. Environ. Plann. B. Plann. Des. **7**(3), 343–351 (1980)

Venselaar, K., Hoop, W, Drunen, P.V.: The Knowledge Base of the designer. In: Simons, P.R.G., Beukh, G. Regulation of Learning, pp. 121–135 (1987)
Wing, J.M.: Computational thinking. Commun. ACM **49**(3), 33–35 (2006)
Wing, J.M.: Computational thinking and thinking about computing. Philosoph. Trans. Royal Soc. A: Math. Phys. Eng. Sci. **366**(1881), 3717–3725 (2008)
Wolfram, S.: Statistical mechanics of cellular automata. Rev. Mod. Phys. **55**(3), 601–644 (1983)

Comparing Human-Centric and Design-Centric Sources: Analysing Pain- Points, Needs, and Requirements Using Pahl and Beitz's Checklist

Gouri Naik[1(✉)], Vishesh Jain[2], Anjan Dhua[2], and V. Srinivasan[1]

[1] Indian Institute of Technology Delhi (IIT Delhi), New Delhi, India
Gouri.Naik@design.iitd.ac.in

[2] All India Institute of Medical Sciences (AIIMS), New Delhi, India

Abstract. This study examines the effectiveness of human-centric versus design-centric sources in identifying pain-points, needs, and requirements that covers more categories of requirement checklists. Using data from a project focused on developing a camera holder for pediatric laparoscopic surgery, the research compares the types of pain-points, needs, and requirements identified through human-centric sources (HCS) and design-centric sources (DCS). Pain-points were gathered by (a) observing surgeries and interviewing surgeons (HCS) and (b) evaluating existing camera holders (DCS). From these pain-points, needs and requirements were subsequently derived. The study reveals that human-centric sources excel in capturing nuanced needs, while design-centric sources address broader pain points. These findings underscore the value of combining both human- and design-centric perspectives to achieve more comprehensive and relevant design outcomes.

1 Introduction

The success of any new product relies on a deep understanding of key elements: the needs and desires of target users, the competitive landscape, and market conditions (Cooper 2003). Brown and Katz (2011) framed this understanding within the context of human-centered design (HCD), where three essential aspects - human desirability, technological feasibility, and market viability - guide the design process. Human desirability focuses on meeting users' needs, feasibility of creating solutions with available technologies or processes, and viability on ensuring designs can succeed in the market.

At the beginning of the design process, designers collect insights as "pain-points" (challenges or frustrations) and "pleasure points" (positive experiences) from a variety of sources, such as stakeholders, existing designs, and market trends. These insights are crucial for defining needs and requirements that guide designing. The information gathered is typically sourced from three sources: human-centric, design-centric, and market-centric sources. The human-centric source (HCS) focuses on understanding the pains, pleasures, needs and experiences of stakeholders—users, customers, and other

K. Gericke et al. (Eds.): DSC 2022/2023, PTT, pp. 197–210, 2026.
https://doi.org/10.1007/978-3-032-16278-6_12

relevant groups—through empathy-based methods like interviews, surveys, observations, and field studies. Incorporating the needs and experiences of stakeholders helps to improve the usability and the overall user experience of the designs (Chammas et al. 2015; Veryzer and Mozota 2005). The design-centric source (DCS) involves analysing existing designs to assess their strengths and weaknesses, which serves as a benchmark for identifying areas for improvement (Hocknell 2019). The market-centric source (MCS), on the other hand, evaluates a design's potential to find acceptance in the market or gain market share, contributing to overall profitability. There is evidence for the benefits of combining HCS and DCS (Kjeldskov and Howard 2004). However, there has been little research to compare the two sources to understand their individual contributions to the design process. This study aims to fill this gap by comparing the HCS and DCS across different stages of product design.

2 Literature Review and Objectives

Comprehensive identification of design requirements is essential to avoid costly project delays and redesigns (Becattini and Cascini 2014). Requirement checklists are widely used to ensure that designers cover all essential considerations, enabling teams to develop high-quality, standardized designs (Altavilla et al. 2022). Checklists such as those by Hales (1993), Pahl and Beitz (1988), and Roozenburg and Eekels (1995) provide frameworks for categorizing design requirements, covering elements like function, safety, ergonomics, cost, and lifecycle factors. These frameworks help designers ensure thoroughness, consistency, and quality in the development process, supporting comprehensive and competitive design outcomes (Becattini et al. 2015).

User-centered design (UCD) emphasizes understanding users' preferences, behaviours, and environments through empathy-driven techniques such as interviews, observations, and focus groups. Studies show that integrating UCD with product lifecycle enhances user satisfaction and product quality but may require significant shifts in organizational culture and resources (Nebe et al. 2006). However, a narrow focus on explicit user feedback can miss underlying and unarticulated needs, and thereby, limiting the depth of insights (Hyysalo, 2004; Shah et al., 2006). For instance, Hyysalo's activity theory emphasized the evolving and mediated nature of user needs, especially in contexts where users encounter unfamiliar technologies. In contrast, DCS focuses on leveraging existing products and design analyses to gain competitive insights. By assessing the strengths and weaknesses of current designs, organizations can identify opportunities for innovation and improvement (Ambaram 2013). Comparative studies have found that user-generated ideas often carry higher novelty and value for customers, while designer-generated ideas may offer greater feasibility (Nishikawa et al. 2013; Poetz and Schreier 2012). Similarly, research by Dahiya and Kumar (2019) highlighted the value of primary user data over secondary data in driving user-centered outcomes. Despite these benefits, DCS can lead to a more product-focused strategy, which may overshadow user insights and result in less user-centered solutions (Cooper 2003; Norman 2005). Integrating HCS and DCS is recognized as valuable for achieving well-rounded design outcomes (Kjeldskov and Howard 2004). While both sources have their strengths and weaknesses, few studies have systematically compared their individual impacts on the design process.

Naik et al. (2025) show that both human-centric (HCS) and design-centric (DCS) sources are equally effective in uncovering critical pain-points and important needs in designing a pediatric laparoscopic camera holder. In another study, Naik and Srinivasan (2024) compared HCS and DCS using a requirements checklist across six student projects, reporting that while both sources covered a similar number of requirement checklist categories, the categories were not the same, underscoring the necessity of integration for comprehensive coverage.

This study aims to investigate the effects of using HCS and DCS independently throughout the design process. Specifically, the research objective is to compare the spanning of categories by pain points, needs, and requirements identified by HCS and DCS. This spanning of categories is assessed using the checklist of requirements developed by Pahl and Beitz (1988). By evaluating the breadth of the coverage of the checklist by pain-points, needs and, requirements, from each source, this study will provide insights into the comprehensiveness and practical utility of HCS versus DCS.

3 Research Methodology

To pursue the research objective, data from a funded design project is used. The aim of the project is to develop and test a laparoscopic camera-holder for paediatric laparoscopic surgery, that is affordable to be used in hospitals in developing countries. The double-diamond model of design thinking of the UK Design Council (2019) is used for designing; it consists of four phases i.e., Discover, Define, Develop, and Deliver. The Discover phase starts with identifying the relevant stakeholders (primary and assistant surgeons) and empathizing with them by observing them during surgeries and interacting with them later. The existing designs of camera-holders are identified and analysed to identify their strengths and weaknesses. These help identify the pleasure- and pain-points of the stakeholders and the existing designs. In the Define phase, needs are elicited from the pain- and pleasure- points and the needs are converted into requirements by quantifying or qualifying some subjective variables. In the Develop phase, conceptual solutions are generated, evaluated (against the needs and requirements), modified (to comply better with the needs and requirements) and the best designs selected for further detailing. For this research, data of pain-points, needs and requirements obtained through the HCS and DCS is used. Although the first author of this paper is involved in this design project, this is not an issue because the design project is carried out before the research goals are formulated. The pain-points are translated into need statements that the camera holder should possess. The need is expressed as an attribute of the design to be developed. Needs are converted into requirements by analysing existing camera holders and taking measurements from the laparoscopic surgical field if required. The pain-points, needs, and requirements identified individually using the human-centric and design-centric sources are categorized using the Pahl and Beitz's checklist. The coverage of the various categories of the checklist by the pain-points, needs and requirements, identified through both these sources is compared.

3.1 HCS

The primary and assistant surgeons are the primary users of the laparoscopic camera-holder. To identify their pain- and pleasure-points, 14 pediatric laparoscopic surgeries at AIIMS, New Delhi, are observed. These surgeries span from 1.5 to 6 h for patients aged 1–12 years. The observations in surgeries are documented through notes and images. These observations are validated through post-surgery discussions and are also used to gain additional insights. Semi-structured interviews with 23 surgeons from six hospitals across India, focusing on challenges and preferences in laparoscopic procedures, are also conducted. These interviews, spanning 1–1.5 h are recorded and transcribed by the first author. The transcriptions are reviewed by the surgeons for their correctness. From the observations of surgeries and interviews of surgeons, the pain-points are identified and clustered based on similarity. From these pain-points, needs and requirements are identified.

3.2 DCS

A review of literature is conducted by the first author to identify various camera-holders used in laparoscopic surgeries. Eighteen models are identified, including FIPS ENDOarm, Naviot, Unitrac® Pneumatic Holding Arm, Autonomous Flexible Endoscope, and ImagTrac (Ma et al. 2019; Mittal et al. 2020; Rahman and Mishra 2011; Rassweiler and Teber 2016; Schneider and Feussner 2017). In the review, each model's features, strengths, and limitations are examined. From these pain-points are discovered, and these are converted into specific needs, which are organized into clusters based on their similarity. The needs are converted into requirements and clustered.

3.3 Requirement Checklist Utilization

The pain-points, needs, and requirements identified through the HCS and DCS are categorized using the 17 categories of the Pahl and Beitz's checklist by the first author. An additional "Other" category is created for entities that do not fit in the categories of the checklist. The definitions of the categories are used in the categorization and are shown in Table 1. To assess the bias in the categorization, an intercoder reliability test using Cohen's Kappa is conducted. Two design researchers, each with at least two years of experience, independently categorize 20% of the identified pain-points, needs, and requirements. This 20% data is selected such that all the categories in the checklist are covered. For this inter-rater reliability test, the researchers are given the checklist, the definitions of the categories, and a few examples of the categorization. The number of categories covered by the pain-points, needs, and requirements from the HCS and DCS are compared.

Table 1. Definition and Acronyms of Pahl & Beitz's Checklist

Category (Acronym)	Definition
Geometry (G)	Refers to physical shape and dimensions of a design, including its size, shape, and orientation; e.g., size, height, breadth, length, diameter, space requirement, number, arrangement, connection, extension, etc
Kinematics (K)	Includes study of motion of a design and type of motion, including its speed, velocity, acceleration, and trajectory without considering external forces
Force (F)	Includes study of forces that act on a design, including the direction of force, the magnitude of force, frequency, weight, load, deformation, stiffness, elasticity, inertia forces, and resonance
Energy (E)	Refers to energy consumption and energy efficiency of a design, including its power consumption and energy output, efficiency, loss, friction, ventilation, state, pressure, temperature, heating, cooling, supply, storage, capacity, and conversion
Material (M)	Includes study of materials used in a design, including physical and chemical properties, auxiliary materials, and prescribed materials along with flow and transport of materials
Signals (Si)	Represent the physical form in which information is channelled; e.g., data stored on a hard drive (information) would be conveyed to the processor via an electrical signal, input & output signals, form, display, etc
Safety (Sa)	Includes safety of a design, including its compliance with safety standards and regulations, including direct safety systems, operational and environmental safety
Ergonomics (Er)	Concerned with designing systems, processes, and products that are comfortable, efficient, and safe for humans to use; e.g., man-machine relationship, type of operation, operating height, clarity of layout, sitting comfort, lighting, etc
Production (P)	Refers to manufacturing process of design, including the equipment, tools, and materials required; e.g., factory limitations, preferred production methods, means of production, achievable quality and tolerances, wastage, etc
Quality Control (QC)	Process of ensuring that a design meets the desired level of quality by identifying defects and taking corrective actions to eliminate them; e.g., possibilities of testing and measuring, application of special regulations and standards
Assembly (A)	Refers to the arrangement and integration of components to create the final product, including considerations for ease of assembly, compatibility, and alignment; e.g., special regulations, installation, sitting, foundations

(*continued*)

Table 1. (*continued*)

Category (Acronym)	Definition
Transport (T)	Includes features and design considerations that facilitate the movement and transportation of a product, including the modes of transportation, logistics involved, means of transport (height and weight), conditions of dispatch, ease of transport and the logistics involved both in its pre-assembled state and during use
Operation (O)	Includes operation of design and operating environment, including its functionality, user interface, and performance; e.g., quietness, wear, special uses, marketing area, destination
Maintenance (Ma)	Refers to maintenance (upkeep and repairs) of a design, including procedures and tools used to maintain the design; e.g., servicing intervals (if any), inspection, exchange and repair, painting, cleaning
Recycling (R)	Includes study related to recyclability of a design, including the materials and processes used in recycling; e.g., reuse, reprocessing, waste disposal storage
Cost (C)	Includes costs associated with a design, including the maximum permissible manufacturing costs, operating costs, maintenance costs, cost of tooling, investment, and depreciation
Schedule (Sc)	Refers to timeline for development, production, and maintenance of a product or system, including the milestones and deadlines involved; e.g., end date of development, project planning, and control, delivery date

4 Results

Table 2 shows examples of pain-points, needs and requirements obtained from the project. The pain points, needs, and requirements identified through HCS and DCS are categorized using the Pahl and Beitz's checklist. Tables 3, 4, and 5 show examples of categorization using the categories of Pahl and Beitz's checklist, for pain points, needs, and requirements, respectively. A total of 6 and 8 high-level pain points were identified through HCS and DCS, respectively. Following the process of high-level categorization, 10 high-level needs emerged from HCS and 8 from DCS. Following high-level categorization, 8 requirements were derived from HCS and 6 from DCS.

Table 2. Examples of pain-points, needs and requirements from project after high-level categorization

Pain-points	Needs	Requirements
1. Stability loss (HCS) 2. Restricted camera manipulation (HCS) 3. Time constraints (HCS) 4. Set-up and assembly of camera-holder is difficult and time-consuming (HCS) 5. Controlling camera-holder imposes cognitive load on surgeons (DCS) 6. Set-up and assembly of camera-holder is difficult and time taking (DCS) 7. Camera-holder is bulky (DCS) 8. Camera-holders not compatible with various kinds of laparoscope (DCS)	1. Camera-holder can be manipulated by both surgeons (primary & assistant)(HCS) 2. Camera-holder is portable (HCS) 3. Accommodates range of motion of the camera needed for performing surgical procedures (HCS) 4. Camera-holder retains its position after reinsertion to provide the same field of vision (HCS) 5. The camera-holder should be a universal camera holder that is compatible with the commonly used laparoscope (DCS) 6. Must not distract or disturb the OT staff, including surgeons (DCS) 7. Camera-holder needs to be orientable in any direction as per the need of the surgery (DCS) 8. The camera-holder should be sturdy (DCS)	1. The camera-holder bears at least 0.880 kg (mass of camera) in any condition (HCS) 2. The camera holder's mass should be less than 7.3 kg if it has to be hand-held and shifted manually (without wheels) (HCS) 3. The camera-holder shall support linear movement with a range of ±10 cm and rotational adjustment of ±30 degrees around the port axis (HCS) 4. The cost of camera-holder is in the range of INR 1–10 lakhs (HCS) 5. The camera-holder should be sturdy enough to withstand a minimum torque of 72Nm(DCS) 6. Setup time should be less than 2 min (DCS) 7. The camera-holder should be a universal camera holder compatible with commonly used laparoscopes such as Olympus CH s400, Stryker 1488, and R.Wolf telescope of 5mm, 10mm, and 12mm diameter (DCS) 8. The camera-holder should provide at least 4 DOF (vertical, lateral, longitudinal, and rotational) (DCS)

Table 3. Examples of Pain-point Categorization

Pain-points	Categorization
Restricted camera manipulation	K, O
The camera-holder is subjected to accidental errors	Sa, QC
Controlling the camera holder imposes a cognitive load on surgeons	Er
The camera-holder is expensive	C

Table 4. Examples of Needs Categorization

Needs	Categorization
The camera-holder is portable	T
The camera-holder provides high positional accuracy	QC, G
The camera-holder is sterilizable	Sa
The camera-holder is easy and quick to operate	Er, O

Table 5. Examples of Requirements Categorization

Requirements	Categorization
The camera holder's mass should be less than 7.3kg if it has to be hand-held and shifted manually (without wheels)	F, T
The maximum permissible dimension for the camera holder should be less than 2000mm(l)*270mm(b)*572mm(h) (Considering Indian anthropometry)	G
The camera holder must be designed to fit within the dimensions of Ø300 mm x 430 mm to ensure compatibility with the chamber size of the Han Sung HTA-30V autoclave	G, Sa

4.1 Result of Intercoder Reliability Test

To assess the extent of agreement between the first author and researchers, individual evaluations are conducted, followed by multiple rounds of intercoder reliability testing. Extensive discussions refine checklist definitions and improve coding clarity. Once a satisfactory Kappa value—indicating high agreement—is reached, the remaining pain-points and needs across all projects are coded. After coding all the pain-points, needs, and requirements, 20% of the data is selected for further analysis. This rigorous approach enhances both the reliability and validity of the coding, strengthening the analysis.

Table 6 presents the intercoder reliability results. In the first round, Kappa values range from 0.562 to 1, reflecting moderate to very strong agreement, as noted by Besar et al. (2012). Following the initial categorization, the first author meets with the researchers to discuss the coding rationale and clarify discrepancies. Differences primarily stem from varying interpretations of checklist categories, such as operation, ergonomics versus safety, and transportation. Based on these discussions, the first author refines category definitions and adds contextual examples to aid understanding. In the second round, Kappa values improve, ranging from 0.785 to 1, indicating strong to very strong agreement.

Table 6. Intercoder reliability assessment conducted in two rounds involving both design researchers and the primary researcher.

	Kappa value between author & 1st researcher		Kappa value between author & 2nd researcher	
	1st Round	2nd Round	1st Round	2nd Round
Pain-points	0.689	0.865	0.927	0.785
Needs	0.762	0.882	1	1
Requirements	0.812	0.878	0.562	0.878

4.2 Comparison of Categories Covered by Human- and Design-Centric Sources

Table 7 provides an overview of the categories covered by pain-points, needs, and requirements identified through HCS and DCS across the project. For pain-points, HCS covers five categories, whereas DCS spans 10 categories. Conversely, needs identified via HCS and DCS cover nine and eight categories respectively in Pahl and Beitz's requirements checklist. For requirements, HCS spans eight categories from the checklist, while DCS covers five categories.

Table 7. Categories covered by human-centric & design-centric sources

Pain-points				Ma	C	Er	K	O	A	G	QC	Sa	T	Si
Needs		T	QC	E	O	A	Er	G	K	Sa	F	Si		
Requirements	M	O	Er	C	F	G	K	Sa	A					

*

Checklist categories covered by HCS
Checklist categories covered by DCS
Checklist categories covered by both HCS and DCS

Figure 1 compares the frequency of the requirement checklist covered by the pain-points gathered from HCS and DCS. It indicates that pain-points derived from DCS generally cover more categories with higher frequencies compared to pain-points derived from HCS.

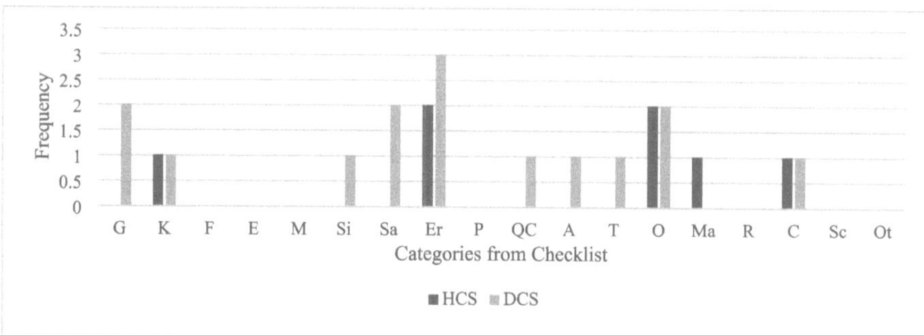

Fig. 1. Frequency of Requirement Checklist Covered by Pain-points from HCS and DCS

Figure 2 shows the frequency of the requirement checklist covered by needs from HCS and DCS. Overall, the figure illustrates that needs derived from HCS generally cover more categories from the checklist with higher frequencies. In contrast, DCS covers certain categories with a lower frequency.

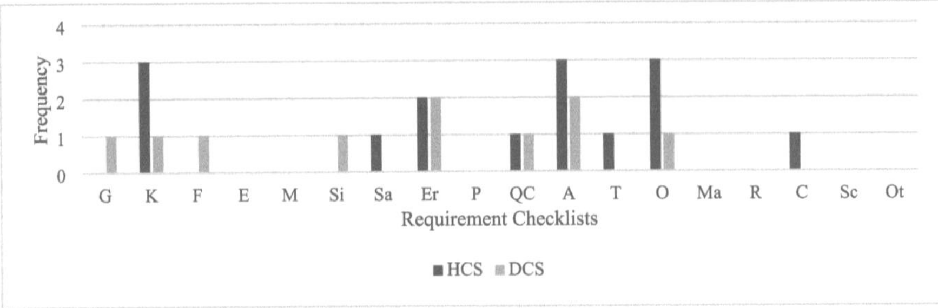

Fig. 2. Frequency of Requirement Checklist Covered by Needs from HCS and DCS

Figure 3 compares the frequency of the requirement checklist covered by requirements gathered from HCS and DCS. The data reveals that requirements derived from HCS generally cover more of Pahl and Beitz's requirement checklist categories and have higher frequencies than DCS.

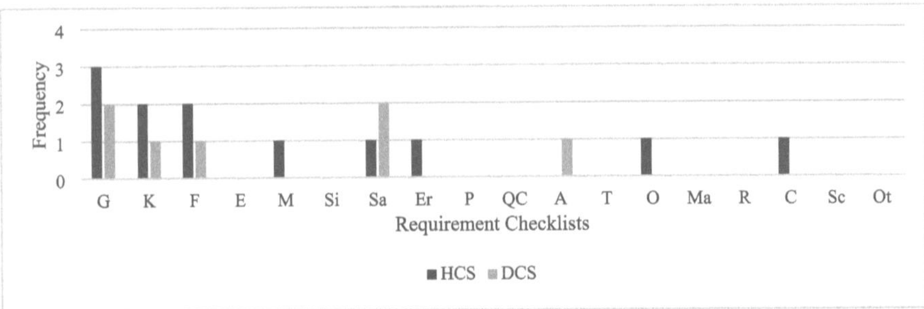

Fig. 3. Frequency of Requirement Checklist Covered by Requirements from HCS and DCS

5 Discussion

This study evaluated whether a human-centric or design-centric approach better captures pain-points, needs, and requirements across a broad set of categories from a requirements checklist. The underlying assumption was that the wider the range of categories covered, the more comprehensive the source of insights. Results showed that the number of categories identified by human- and design-centric source covered distinct categories. This suggests that using both sources together could provide a more complete picture, encompassing more checklist categories than either source alone. For instance, pain-points identified through human-centric methods covered categories like Material, Cost, Ergonomics, Kinematics, and Operation, while those from design-centric methods covered Cost, Ergonomics, Kinematics, Operation, Assembly, Geometry, Quality Control, Safety, and Transportation. Together, these approaches could collectively address categories such as Geometry, Ergonomics, Material, Maintenance, Operation, Safety, Kinematics, and Cost more comprehensively.

The pain points derived from DCS cover a broader range of categories (10) compared to those from HCS (5). This finding suggests that DCS has a more extensive scope in identifying diverse pain points. For instance, the Ergonomics (Er) category appears more frequently in DCS (3 times) than in HCS (2 times). Additionally, DCS exclusively addresses categories such as Geometry (G), Safety (Sa), Assembly (A), Quality Control (QC), Signal (Si), and Transportation (T), although with varying frequencies. Conversely, HCS uniquely covers the Maintenance (Ma) category, highlighting its strength in addressing this aspect, which DCS overlooks. Interestingly, certain categories, including Cost (C), Kinematics (K), and Operation (O), are equally covered by both HCS and DCS, demonstrating a convergence in addressing these aspects. Despite this overlap, the higher frequency of coverage for many categories in DCS underscores its dominance in addressing pain points, as visualized in Fig. 1.

The needs derived from HCS show broader and more frequent coverage of Pahl and Beitz's checklist compared to those from DCS. Figure 2 highlights that HCS covers critical categories such as Kinematics and Operation more extensively (3 times) than DCS (once). Additionally, HCS addresses Geometry, Force, and Signal categories with higher frequencies, while DCS does not cover them. The Ergonomics category is covered equally by both sources, suggesting parity in this specific area. Notably, HCS uniquely addresses some categories with higher frequencies, such as Assembly and Kinematics, emphasizing its strength in capturing nuanced human-centered needs.

Requirements derived from HCS exhibit broader and higher-frequency coverage of Pahl and Beitz's checklist categories than those from DCS, as depicted in Fig. 3. For example, HCS significantly outperforms DCS in categories like Geometry (G), Kinematics (K), and Force (F), covering these categories more than twice as often as DCS. On the other hand, DCS provides notable contributions to categories such as Safety (Sa) and Geometry (G), covering these twice. Overall, HCS demonstrates a stronger capacity to address a wider array of requirements comprehensively. Its focus on high-frequency coverage in Geometry, Kinematics and Force categories further underscores its suitability for capturing detailed and diverse requirements. While DCS has strengths in addressing specific categories with notable frequency, its overall coverage is comparatively limited.

The findings of this study are subject to several limitations. For the human-centric source (HCS), pain points and needs were identified through observations of multiple surgeries and interviews with primary and assistant surgeons, with verification from relevant surgeons. In contrast, the design-centric source (DCS) relied on a review of existing literature to identify pain points and needs. A notable limitation is the absence of direct exposure to or observation of available camera-holder designs, as these are predominantly found in Western markets. Additionally, data from the HCS were exclusively derived from end users—specifically, primary and assistant surgeons—without input from other stakeholders, such as manufacturers, camera developers, or hospital administrators. While the study's goal was to explore the influence of human-centric and design-centric sources on design outcomes throughout product development, it focused on a single case: the design of a laparoscopic camera-holder. Despite the variety of users consulted and existing designs analyzed, the camera-holder represents only one data point. The applicability of HCS and DCS may vary significantly depending on the domain or specific design challenges, with some contexts requiring a greater emphasis on one source over the other. Moreover, the exclusion of secondary and tertiary stakeholders limits the study's insights into how broader stakeholder involvement might influence outcomes. This deliberate focus on primary users constrained the scope of the findings. The research aimed to evaluate the impact of HCS and DCS on design outcomes and requirements coverage across design stages. Future work could explore the relative importance of pain points and needs derived from each source. Current findings suggest that integrating both sources may provide the most comprehensive design insights. Expanding comparisons across diverse domains and design contexts will be critical for future research.

6 Summary, Conclusion and Future Work

The goal of this research was to explore the distinctions between human-centric and design-centric sources by examining the extent to which each source addresses various categories in a requirements checklist through the pain-points, needs, and requirements identified from each. Using data from a design project focused on a camera holder for pediatric laparoscopic surgery, the study collected pain-points from human-centric sources (HCS) by observing surgeries and interviewing surgeons, and from design-centric sources (DCS) by evaluating existing camera holders. The findings highlight distinct strengths of HCS and DCS in identifying pain points, needs, and requirements. HCS excels in capturing a broader and more nuanced spectrum of needs and requirements, emphasizing its value in human-centric design processes. DCS, on the other hand, demonstrates a broader scope in addressing pain points, suggesting its utility in uncovering diverse design-related challenges. These complementary strengths suggest that integrating both sources could enhance the design process by leveraging the comprehensive insights from HCS and the broader pain point identification from DCS. This approach would ensure a balanced consideration of human and design-centric factors, leading to more robust and effective solutions. This study is limited by its reliance on observations and interviews with primary users (surgeons) while excluding other stakeholders and by focusing on a single case, the laparoscopic camera-holder. The findings may not generalize across domains, emphasizing the need for broader comparisons and stakeholder inclusion. Despite these limitations, the research provides valuable insights into how these distinct sources shape design outcomes and strongly supports integrating human-centric and design-centric approaches to create a more comprehensive design process. These findings align with related studies that underscore the complementary nature of both sources, advocating for their combined use in design practices to achieve more thorough and effective solutions.

References

Altavilla, S., Becattini, N., Fiorineschi, L., Rotini, F.: Effectiveness of different requirements checklists for novice designers. J. Integr. Des. Process. Sci. **26**(1), 45–69 (2022). https://doi.org/10.3233/JID-210015

Ambaram, M.R.: The factors that enable customer centricity and the changes in the organisation design when moving from a product to a customer centric strategy, Doctoral dissertation, University of Pretoria (2013). https://doi.org/10.2116/bunsekikagaku.61.637

Becattini, N., Cascini, G.: General-purpose requirements checklist for improving the completeness of a design specification. In: Proceedings of International Design Conference, DESIGN, Vol. 2014-Janua, pp. 111–120 (2014)
Becattini, N., Cascini, G., Rotini, F.: Requirements checklists: benchmarking the comprehensiveness of the design specification. In: Proceedings of the International Conference on Engineering Design, ICED, vol. 5 No. DS 80-05, pp. 1–10 (2015)
Besar, M.N.A., et al.: Should a single clinician examiner be used in objective structure clinical examination?, Procedia - Social and Behavioral Sciences, vol. 60 No. October, pp. 443–449 (2021). https://doi.org/10.1016/j.sbspro.2012.09.404
Brown and Katz, 2011. Brown, T., Katz, B.: Change by design. J. Product Innov. Manage. Wiley **28**(3), 381–383 (2011). https://doi.org/10.1111/J.1540-5885.2011.00806.X
Chammas, A., Quaresma, M., Mont'Alvão, C.: a closer look on the user centred design. Procedia Manufacturing, Elsevier B.V., vol. 3 No. December, pp. 5397–5404 (2015). https://doi.org/10.1016/j.promfg.2015.07.656
Cooper, R.G.: Profitable product innovation: the critical success factors. The International Handbook on Innovation, pp. 139–157 (2003). https://doi.org/10.1016/B978-008044198-6/50010-3
Dahiya, A., Kumar, J.: Do design outcomes get influenced by type of user data? an experimental study with primary and secondary user research data. In: Ahram, T., Karwowski, W., Taiar, R. (eds.) Human Systems Engineering and Design. IHSED 2018. Advances in Intelligent Systems and Computing, vol. 876. Springer, Cham (2019). https://doi.org/10.1007/978-3-030-02053-8_30
Hales, C.: Managing Engineering Design, Longman Scientific {&} Technical (1993)
Hocknell, M.: Profit by design: how to build a customer portfolio full of profitable promoters. Hambone Publishing (2019)
Hyysalo, S.: Users, an emerging human resource for R&D? From eliciting to exploring users' needs. Int. J. Hum. Resources Develop. Manage. **4**(1), 22–37 (2004). https://doi.org/10.1504/IJHRDM.2004.004491
Kjeldskov, J., Howard, S.: Envisioning mobile information services : combining user- and technology-centered design. In: Masoodian, M., Jones, S., Rogers, B. (eds.) Computer Human Interaction. APCHI 2004. LNCS, vol. 3101. Springer, Berlin, Heidelberg (2004). https://doi.org/10.1007/978-3-540-27795-8_19
Ma, X., Song, C., Chiu, P.W., Li, Z.: Autonomous flexible endoscope for minimally invasive surgery with enhanced safety. IEEE Robot. Autom. Let. **4**(3), 2607–2613 (2019). https://doi.org/10.1109/LRA.2019.2895273
Mittal, R., Sbaih, M., Motson, R.W., Arulampalam, T.: Use of a robotic camera holder (FreeHand®) for laparoscopic appendicectomy. Minimally Invasive Therapy and Allied Technologies, Taylor & Francis, vol. 29, no. 1, pp. 56–60 (2020), https://doi.org/10.1080/13645706.2019.1576052
Naik, G., Jain, V., Dhua, A., Srinivasan, V.: Comparison of human-and design-centric sources: discovery of pain-points and identification of needs. Respon. Resilient Des. Soc. (2025). https://doi.org/10.1007/978-981-96-7308-7
Naik, G., Srinivasan, V.: Human- and design-centric source: comparison using requirements checklist. Proc. Des. Soc. **4**, 65–74 (2024). https://doi.org/10.1017/pds.2024.9
Nebe, K., Grötzbach, L., Hartwig, R.: Integrating user centered design in a product development lifecycle process: a case study. In: Proceedings of 2006 International Conference on Software Engineering Research & Practice. CSREA Press, Las Vegas, S, No. February 2014, pp. 695–701 (2006)
Nishikawa, H., Schreier, M., Ogawa, S.: User-generated versus designer-generated products: a performance assessment at Muji. Int. J. Res. Mark. Elsevier B.V. **30**(2), 160–167 (2013). https://doi.org/10.1016/j.ijresmar.2012.09.002

Norman, D.A.: Human-centered design considered harmful. Interactions **12**(4), 14–19 (2005). https://doi.org/10.1145/1070960.1070976

Pahl, G., Beitz, W.: Engineering design: a systematic approach, Nasa Sti/Recon Technical Report A, vol. 89 (1988)

Poetz, M.K., Schreier, M.: The value of crowdsourcing: can users really compete with professionals in generating new product ideas? J. Prod. Innov. Manag. **29**(2), 245–256 (2012). https://doi.org/10.1111/j.1540-5885.2011.00893.x

Rahman, M.N., Mishra, R.K.: The camera-holding robotic device in laparoscopy surgery. World J. Laparoscopic Surg. **4**(3), 132–135 (2011). https://doi.org/10.5005/jp-journals-10007-1130

Rassweiler, J.J., Teber, D.: Advances in laparoscopic surgery in urology. Nat. Rev. Urol. **13**(7), 387–399 (2016). https://doi.org/10.1038/nrurol.2016.70

Roozenburg, N.F.M., Eekels, J.: Product design: fundamentals and methods. Wiley **24** (1995). https://doi.org/10.1016/s0925-7535(97)81485-3

Schneider, A., Feussner, H.: Mechatronic support systems and robots, biomedical engineering in gastrointestinal surgery, pp. 387–441. Academic Press (2017). https://doi.org/10.1016/b978-0-12-803230-5.00010-5

Shah, D., Rust, R.T., Parasuraman, A., Staelin, R., Day, G.S.: The path to customer centricity. J. Serv. Res. **9**(2), 113–124 (2006). https://doi.org/10.1177/1094670506294666

UK Design Council. Framework for innovation: design council's evolved double diamond - design council, 17 May (2019). https://www.designcouncil.org.uk/our-work/skills-learning/tools-frameworks/framework-for-innovation-design-councils-evolved-double-diamond/. Accessed 10 March 2023

Veryzer, R.W., Mozota, B.B.D.: The impact of user-oriented design on new product development: an examination of fundamental relationships. J. Prod. Innov. Manag. **22**(2), 128–143 (2005)

Methodological Implications of Coevolution of Design Thinking with Systems Thinking and Computational Thinking

Vishal Singh(✉)

Indian Institute of Science, Bengaluru, India
singhv@iisc.ac.in

Abstract. Design thinking has been extensively discussed in recent years. It is well-established that design thinking is an iterative process that involves co-evolution of problem and solution. In parallel to design thinking, there is growing interest in systems thinking and computational thinking, driven by the recognition of system-level dependencies in design and the explicitly growing use of computational methods in problem-solving. While design thinking, systems thinking and computational thinking have typically been discussed separately, researchers have recently observed and highlighted the interaction between these three thinking paradigms. Accordingly, this research theoretically investigates the iterative flow between design thinking, systems thinking, and computational thinking. The article argues that these three types of thinking are intertwined, and they co-evolve during the design process. The methodological implications of this coevolution are discussed.

1 Introduction

Design thinking, as a cognitive process to foster creative exploration of the design space, has remained a core area of research interest among design researchers, design educationists, and design practitioners (Lawson 2006, Gero 1990, Schon 1992, Brown 2008, Cross 2011, Dorst 2011). Over the years, multiple design thinking frameworks have emerged in the design research community, building on a heuristics-based approach, following the principles of bounded rationality and satisficing (Simon 1969, 1982). Consequently, all design thinking frameworks implicitly build on the premise that: (1) all creative design problems have more than one satisfactory solution that is sufficiently good, and (2) designers will have to contend with a partial understanding of the problem at the beginning of the design process and seek further understanding of the problem as the design progresses. Following this premise, research into design thinking has led to several notable insights into various aspects of design process, including problem-solution coevolution (Lawson 2006, Schon 1992, Poon and Maher 1997), iterative divergence and convergence of the problem and solution spaces (Cross et al 1996, Dorst and Cross 2001), iterative problem formulation and reformulation (Gero 1990), and the critical role of framing and abductive reasoning in the design process (Dorst 2011). These insights, the resulting impact, and simplified versions of the design thinking frameworks

K. Gericke et al. (Eds.): DSC 2022/2023, PTT, pp. 211–222, 2026.
https://doi.org/10.1007/978-3-032-16278-6_13

have garnered widespread attention, leading to the popularization of design thinking across multiple domains (Brown 2008), extending beyond the design community. Consequently, the application of design thinking has been reported from all kinds of problem contexts and scenarios, including the design of physical and digital products, services, educational programs, governmental policies, organisations, product-service systems, etc.

In contrast, systems thinking (Arnold and Wade 2015, Forrester 1987, Meadows 2015) and computational thinking (Wing 2006, 2008) have remained niche and restricted to specialised domains of applications. While systems thinking is a problem-solving approach that views issues within the context of a larger, dynamic system, considering the interconnectedness and interdependencies of its components (Arnold and Wade 2015), computational thinking is a problem-solving approach that involves breaking down problems, identifying patterns, abstracting details, and creating step-by-step instructions to solve them (Wing 2006, 2008).

Even within the design domain, there has been limited discussion and proliferation of systems thinking and computational thinking, both within research and practice. However, more recently, this is changing. Some researchers have brought to attention the intertwining of design thinking with systems and computational thinking (Shute et al 2017, Kelly and Gero 2021, Singh 2025). For instance, Singh (2025) has argued that design thinking is not feasible without systems and computational thinking. Similarly, the growing number of systems-thinking-related courses and programs in leading design institutes is a testament to the growing recognition that systems thinking needs to be an integral part of design education. While computational thinking is still to be adequately integrated into design education, the push for computational thinking in school education as a primary skill, alongside design thinking as a primary skill, is a positive trend. This chapter builds on this trend. In particular, building on Singh (2025) and others (e.g., Shute et al 2017, Kelly and Gero 2021), this chapter argues that systems thinking and computational thinking are integral to design thinking. The chapter aims to argue and demonstrate that design thinking, systems thinking, and computational thinking co-evolve during the design process.

The rest of the chapter is organised into five sections. Section 2 presents a brief review of the literature on the three thinking paradigms. Section 3 presents the theoretical approach adopted in this article to establish the implicit co-evolution of the three thinking paradigms during design. This approach is built around a fundamental review of some of the common definitions of design as a basis to revisit the design thinking frameworks. Section 4 synthesises the arguments based on the reviews in Sects. 2 and 3. and discusses the methodological implications of the iterative dependence between the three thinking paradigms, which are followed by Conclusions.

2 Literature

2.1 Design Thinking

Simon (1969, 1982) outlined the importance of concepts like heuristics, satisficing, and bounded rationality in solving real-world problems. Heuristics are simple rules of thumb for rendering a judgment or making a decision. The word 'satisficing' was coined by combining satisfaction and sufficient to argue and emphasise a heuristics-based approach to seeking good enough or acceptable solutions in a problem context where finding an optimal solution is either unlikely or requires unreasonable time and resources. The bounded rationality of a decision maker is the limits based on the decision maker's knowledge at the time of decision-making, their thinking capacity, the information available at that time about the problem, and the sequence of events leading to the problem situation. Since Simon's 'Sciences of the Artificial' was published, several design researchers have studied human designers and design process, aiming to understand and explain the processes in design thinking.

Multiple researchers have argued that design problems are ill-defined (Simon 1982, Gero 1990), where the problem context is only partially understood at the beginning of the design process. This has led to the recognition that designing involves a reflective practice (Schön 1992), and the understanding of problem and solution coevolves during the design process (Poon and Maher 1997, Dorst and Cross 2001). Several notable frameworks and insights have been reported around design thinking, often explained using different terminologies. The first notable framework evolved from functional reasoning and functional synthesis, leading to Gero's design prototype (1990) that explained the iterative design process using the Function-Behaviour-Structure (FBS) ontology. The second notable framework, which partly emerged from the discontent with the apparent inadequacies of Gero's FBS ontology and the functional reasoning approach, emphasised the importance of framing and abductive reasoning in design process (Dorst 2011). Dorst argued that, unlike other problems that can be solved completely with inductive or deductive reasoning, ill-defined design problems with partial information to begin with typically require designers to apply abductive reasoning, where designers need to hypothesise and explore a potential path of solution approach that is likely to lead to a satisficing solution. Dorst (2011, 2015) distinguishes between normal abduction and design abduction to emphasize the challenge of dealing with the ill-defined nature of design problems. While the first two notable design thinking frameworks remain popular and highly cited in the academic literature, the Standford d-School Design Thinking framework (Brown 2008) truly popularized and disseminated design thinking in industry and practice, both within and outside the design domains. While the d-school Design Thinking framework also highlighted the iterative nature of design, noting the importance of iterative prototyping and testing, it stands out from the other design thinking frameworks for emphasising a human-centred approach grounded in empathy. Hence, a brief overview and comparison of the three design thinking frameworks is presented.

Gero's (1990) design prototype explains the iterative design process around the FBS ontology, revolving around only three key constructs. Besides decomposing the design artefact in terms of the FBS constructs, it also decomposes the design process into eight discrete steps, including three kinds of problem reformulations. The notable aspect of

Gero's design thinking framework is its recursive nature. The same loop of design reasoning can be applied to a design problem at any level of detail. Thus, Gero's articulation of the recursive design reasoning should theoretically be adequate to provide an algorithm for computational implementation. However, what is not clearly articulated is how to hierarchically decompose the problem at the different levels of functional decomposition. More generally, the broader criticism of Gero's design prototype is associated with the ambiguity in the definitions of functions, behaviours and structures, especially confusion between functions and behaviours and between behaviours and structural attributes.

Dorst (2011, 2015) differentiates design problems from other kinds of problems by emphasising the importance of framing and abductive reasoning, the two aspects that are the core of design thinking from his perspective. Dorst argues that problem-solving can be represented as a system of equations involving 'what', 'how', and the desired 'outcomes' (see Fig. 1). According to Dorst, design abduction is the most challenging problem because both 'what' and 'how' are unknown at the beginning of problem-solving, while the outcomes are also yet to be determined. With the help of theoretical arguments supported by examples, Dorst illustrates how framing can help the designers reduce the number of unknowns in design abduction by leveraging abductive reasoning to adopt a working hypothesis of 'how' a problem could be solved.

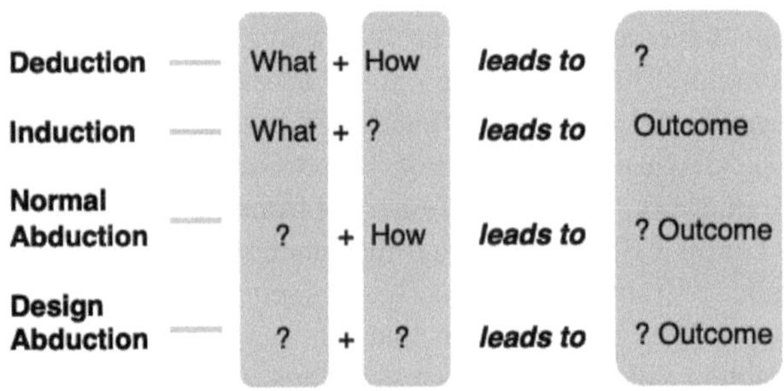

Fig. 1. Framing and design abduction as the core of design thinking (Based on Dorst 2011)

As noted earlier, the Stanford d-school model of Design Thinking (Brown 2008) popularised design thinking frameworks in practice across different domains. One of the most significant contributions of the d-school Design Thinking framework is emphasizing the problem understanding phase, especially empathizing with stakeholders in the context, targeting a human-centred design. The d-school Design Thinking framework succeeded in bringing attention to human aspects of design, which often gets overlooked by technologists and engineers. As a result, the d-school Design Thinking framework also highlighted the importance of user interaction and user experience. From the perspective of creating awareness, another key contribution of the d-school Design Thinking framework was the ability to bring attention to the importance of problem research and discovery, convergence and divergence, etc., reiterating the double diamond representation. While the importance of problem formulation, problem-solution coevolution, convergence and divergence, exploration and exploitation, and iterative prototyping and testing has been emphasized time and again in the design literature, the d-School Design Thinking framework succeeded in creating a simple and effective visual representation of these aspects of design thinking to the wider audience.

2.2 Computational Thinking

In the early 2000s, Jeannette Wing (2006, 2008) used and popularized the term Computational Thinking, especially in computer science and education. The discussions around computational thinking have not only focused on how computer scientists think and act; computational thinking is also viewed as a fundamental approach to how humans think and compute, with or without using computers.

Wing (2006, 2008) broadly classified four ways of thinking and computing to resolve a problem: decomposition, abstraction and representation, algorithms and steps, and pattern recognition. Decomposition allows solving complex problems by reducing them into smaller, manageable problems. Abstraction allows solving a complex problem by removing unnecessary details and reducing the problem into a simplified form that is easier to resolve. Multiple authors have claimed that abstraction is the most critical aspect of computational thinking. Algorithms and steps facilitate problem-solving by outlining a process and decision flow that allows different aspects of the problem to be resolved at different stages of problem-solving, thereby reducing the complexity of managing all aspects of the problem simultaneously. And finally, one of the key aspects of any problem-solving is the ability to recognize patterns. Pattern recognition is an analytical requirement during the problem formulation and abstraction to decide what is relevant and what is not. At the same time, pattern recognition is required to analyse and evaluate outcomes. Pattern recognition can be deductive, inductive, or abductive. For humans, pattern recognition is an innate ability that allows us to learn from our experiences. Not surprisingly, pattern recognition remains a core focus in the computational domain and machine learning.

Other aspects of computational thinking that have been highlighted by some of the researchers include debugging, iteration, and generalisation. Debugging is a critical step in software development. However, the underlying need for debugging should be observable in all kinds of computational activities that require testing, evaluation, and fixing the solution. Typically, this debugging is also a part of the overall problem-solution iteration. Conceptually, this is like the prototyping-testing-fixing-ideate iteration cycle in design thinking. The generalisation aspect of computational thinking highlights the need for methods, processes and solutions that can be applied to multiple cases, typically to find an approach that applies to a class of problems. That is, instead of finding an instance-specific solution, generalisation seeks a solution approach that can be repeated and applied to similar problems.

2.3 Systems Thinking

Systems thinking views a problem from the perspective of the parts and the whole, and the relationships and interdependencies between them. System thinking builds on the premise that no system functions in isolation. Hence, the interactions and interdependencies between the parts of the system are critical.

Systems thinking is discussed and studied in various forms, such as systems engineering, systems dynamics, systems modelling, systems archetypes, and ecosystems.

One of the fundamental steps in systems thinking is the recognition and requirement to assess the system architecture, including the elements and parts of the system, the links and connections between the elements and parts, and the interdependencies between the elements and parts of the system. The interactions between the parts of the system can lead to the emergence of new possibilities and behaviours in the system which cannot be observed otherwise (Forrester 1987). Such emergent system-level behaviours demonstrate that the whole is more than the sum of the parts of the system.

Another critical step in applying systems thinking is determining the system boundary– what is part of the system and what is not (Forrester 1987, Meadows 2015). All the elements and parts not within the system must be considered part of the environment. Systems can be classified as open, closed, or isolated systems, depending on the system boundary and the system's interaction with its environment. Along the same lines, the Theory of Technical Systems (Hubka and Eder 1988) discusses the input-transformation-output through the system boundary regarding the flow of information, material, energy, and resources. Similarly, systems thinking has been applied to understand business ecosystems (Moore 1996), ecological systems (Holling 1973), and innovation ecosystems (Granstrand and Holgersson 2020), similarly focus on understanding the interdependencies in a system to assess the likely feasibility or impact of a potential intervention or change in one part of the system.

The interactions and interdependencies between the elements and parts of the system determine how the changes in one part affect the other parts and the overall system behaviour. This assessment of change propagation in a system is called system dynamics (Forrester 1987, Meadows 2015). System dynamics includes the identification and modelling of the causal loops within the system. The causal loops can be qualitatively mapped to show which elements are positively or negatively impacted by a change in another connected part of the system. Besides the qualitative understanding of the positive and negative loops, it is possible to quantitatively understand and assess the causal influences in terms of the rate of change in influence, which can be parametrized and calculated in terms of stocks, flows and rates.

2.4 Literature Linking Design Thinking, Computational Thinking and Systems Thinking

More recently, researchers have started investigating the relationship between design thinking, computational thinking, and systems thinking. For instance, Kelly and Gero (2021) proposed a dual-process model of design thinking and computational thinking based on the orientation of the thinker towards the generalizability of problems and solutions. From the perspective of problem framing, Kelly and Gero argue that design thinking seeks the generalisation of problem framing, while computational thinking seeks specific framing. From the perspective of solution outcome, Kelly and Gero (2021) argue that design thinking seeks unique solutions, while computational thinking seeks generalisable solutions. However, these arguments are questionable. There is no reason to assume that computational thinking could not be used to explore and generalize the problem framing. Similarly, there is no reason to assume that design thinking could not be used to seek generalizable solutions.

Shute et al (2017) argue that computational thinking encompasses both design thinking and systems thinking. They argue that system thinking allows thinkers to identify system boundaries, model, represent and simulate the system's working, and make decisions about the model. In contrast, computational thinking focuses on the workings of the whole system. They argue that computational thinking focuses on practical and theoretical problems, whereas design thinking primarily focuses on practical problems. This claim is problematic and questionable. There is no reason to assume that design thinking only focuses on practical problems or cannot be applied to theoretical problems.

More research is needed to not only understand the differences between the three thinking paradigms but also to find synergies and interactions between the three. From the perspective of design research, such an investigation should start from the fundamental review of design itself. The question is – what aspects of design, and consequently design thinking, reflect fundamental relationships to computational thinking and systems thinking?

3 Theoretical Approach– Grounded in a Fundamental Review of Design Definitions

To theoretically investigate the interrelationships between the three thinking paradigms, we review some notable definitions and statements on design coming from different perspectives.

From the perspective of a design practitioner, Charles Eames states that "Design is a plan for arranging elements in such a way as best to accomplish a particular purpose." Accordingly, some of the fundamental courses in almost all arts and design schools include elements and principles of design. A design is a composition (whole) of elements (parts). The principles of design pertain to how the elements are interrelated or arranged with respect to each other. Thus, a systems view is implicitly embedded in this definition of design, referring to the elements (parts), their arrangement (interdependence), and the overall composition (whole). At the same time, the definition implicitly reflects aspects of computational thinking by referring to (de)composition–elements, pattern recognition and algorithms– through assessment of arrangement and whether the purpose is accomplished or not. That is, it is implicit within this definition that the designer would need to develop alternative solutions with different arrangements of elements and ensure through evaluation that the intended purpose or goal is accomplished.

From the perspective of design research and design cognition, Gero (1990) states that "design is purposeful and the activity of designing is goal-oriented. The meta-goal of design is to transform requirements, more generally termed functions which embody the expectations of the purposes of the resulting artefact, into design descriptions." Gero's definition refers to the transformation of requirements into design descriptions, including a process of evaluation that ensures that purposes and goals are met. Once again, this definition refers to aspects of systems thinking and computational thinking, including a system comprising inputs, transformation, and outputs. Furthermore, it is important to note that Gero's definition of design is accompanied by his description of the design prototype and the FBS ontology. Thus, the system is decomposed in terms of the FBS constructs. The relationships between these FBS constructs are defined in terms of the causal relationships between these constructs. From a computational thinking perspective, the design prototype defines the recursive algorithm.

From the perspective of problem-solving and decision-making, Simon states that "Everyone designs who devises courses of action aimed at changing existing situations into preferred ones." Thus, Simon argued a broader perspective of design and its scope. Further, Simon's discussion on design built on the concepts of satisficing and bounded rationality. Accordingly, over the years, Simon's definition and articulation of design have been interpreted as "a plan for change from an undesired state to a desired state." This definition subsumes multiple aspects that reflect systems thinking and computational thinking. First, it brings the notion of state change and assessment or evaluation of the state. Second, it brings to attention the plan, which involves a sequence of steps or algorithms for transitioning from one state to another. Third, it brings to attention the need to assess patterns in context to understand whether the state is desired or not. Fourth, the assessment of desire, combined with the concept of satisficing, becomes contextual and situational.

Table 1 presents a summary of the systems thinking and computational thinking aspects implicit in the different definitions and descriptions of design.

Table 1. Systems thinking and computational thinking aspects implicit in design definitions

Design definition/description	Systems Thinking	Computational Thinking
Design is a plan for arranging elements in such a way as best to accomplish a particular purpose- Charles Eames Related: elements and principles of design	Elements (Parts), Design/Composition (Whole) Way of arranging elements (Interdependence)	Elements (Decomposition) Way of arranging (Patterns, steps) Plan (sequence of steps/ representation)
Design is purposeful and the activity of designing is goal-oriented. The meta-goal of design is to transform… into design descriptions Related: FBS ontology, design prototype	Structures (Parts) Design description (whole) Design prototype (relationships between FBS)	FBS (Decomposition) Design prototype, transform (Steps, algorithms)
… courses of action aimed at changing existing situations (undesired state) into preferred ones (desired state) Related: states, satisficing, bounded rationality	Elements with states (Parts) Action and change (System with boundary for input-transformation-output) Action and states (Interdependencies and relationships)	States (Decomposition) Situations (Patterns) Course of action (Steps, algorithms)

3.1 Synthesis of the Theoretical Findings – Interaction Between the Three Thinking Paradigms

Figure 2 shows the overlaps and complementarities between design thinking, computational thinking, and systems thinking.

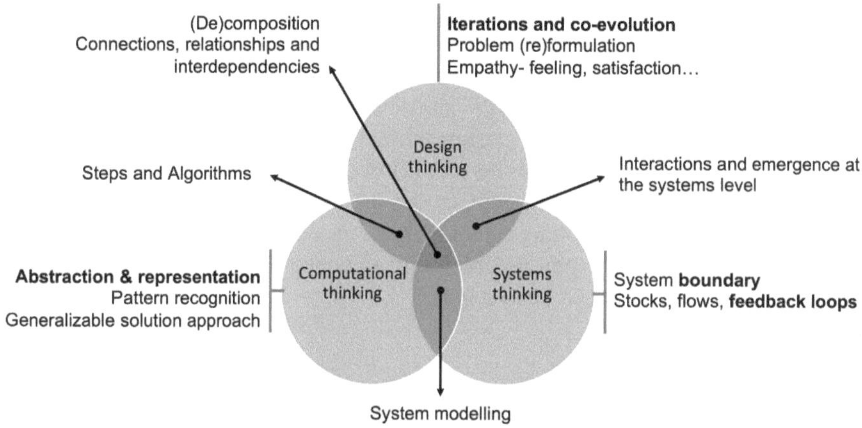

Fig. 2. Relationships between design thinking, computational thinking, and systems thinking

Overlaps Between Design Thinking and Computational Thinking: Some aspects common between design thinking and computational thinking include abstraction and problem formulation, pattern recognition, steps and algorithms, and iterations. Though abstraction remains implicit in design thinking, the problem formulation step requires the ability to prune unnecessary details and focus on what needs to be solved. This has been reinforced in the double diamond framework, which explicitly highlights the importance of finding the right problem to solve. Further, to figure out what problem to solve, the designer must recognize and assess the patterns in the existing situation to identify where the gaps are. Similarly, design thinking also involves steps and algorithms, as noted in Gero's (1990) design prototype or d-school's iterative Design Thinking framework or Dorst's (2011, 2015) framing and abductive reasoning. For instance, Gero's design prototype (1990) includes the iterative steps of requirements, problem (re)formulation, synthesis, analysis, evaluation, and documentation. Though iteration is explicitly discussed in design thinking, the computational thinking frameworks do not explicitly refer to iteration. Some variants of the computational thinking framework point towards iterations, including debugging and iterative prototyping, testing, debugging, and refinement, which has led to methodologies such as agile development.

Overlaps Between Design Thinking and Systems Thinking: Some aspects common between design thinking and systems thinking include elements (parts) and composition (whole) with interdependencies, connections and relationships, and system boundaries. From the perspective of problem-solving, as discussed by Simon (1969, 1982), the need for a design intervention comes from the desire for change within an existing system that is undesirable. Second, for a design solution to satisfy the purpose within a context, it is imperative to understand the system boundary, as well as the interfaces and interdependencies within and outside the system. This overlap between design thinking and systems thinking, including recognising the system boundary and corresponding input-transformation-output, is explicitly observable in the Theory of Technical Systems (Hubka and Eder 1988). Similarly, Dorst's (2011, 2015) description of design thinking requires identifying system components– 'what', the relationships and interdependencies– 'how', and the whole– the design outcome. Gero's (1990) design prototype requires identifying the system components– Structures, the relationships and interdependencies- Behaviours and functions, and the whole– the design output.

Reviewing the overlap between design thinking and systems thinking also helps theoretically resolve some of the gaps between Dorst's (2011) and Gero's (1990) frameworks on design thinking. For instance, Gero's FBS constructs can also be represented using Dorst's framing and reasoning equation using what, how, and output descriptions. In addition, Dorst's equation can also be used to connect elements and principles of design to the design outcome, as shown in Fig. 3.

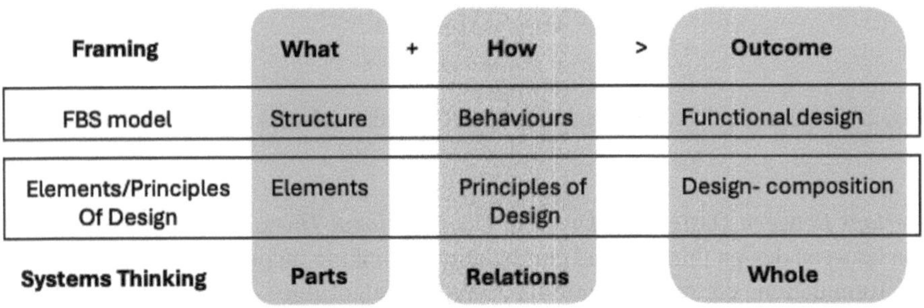

Fig. 3. Superposing framing in design thinking with systems thinking viewpoint

Overlaps Between Computational Thinking and Systems Thinking: Some aspects common between computational thinking and systems thinking include decomposition, part-whole relationships, connections and interdependencies, and systems modelling and representation.

To apply computational thinking, one must start by identifying the system to compute about. The computational processes of abstractions or decomposition are intertwined with the systems view of identifying part-whole relationships and their connections and interdependencies. To consider the input-transformation-output relationships, it is imperative to determine and consider the system boundary.

Insights into the Complementarities and Unique Value Adds from Each Thinking Paradigm: Each of the three thinking paradigms provides unique insights and viewpoints.

Design thinking brings specific attention to the iterative coevolution of the understanding of problems and solutions. This highlights the importance of initiating problem-solving despite partial information about the problem context. To navigate such a situation with partial information, framing and design abduction (Dorst 2011) form the core of design thinking. Similarly, design thinking highlights the need for openness and willingness to allow problem reformulation if required. The emphasis on empathy and human-centred design is also unique to design thinking. This is critical because satisficing is eventually a human judgement.

Systems thinking brings specific attention to system boundaries. Similarly, an understanding of system dynamics is implicit and necessary for assessing system-level feasibility and the potential impact of design interventions in the context. The context defines the system within which the design is expected to function.

Computational thinking brings attention to abstraction, pattern recognition, and computable representation. Computational thinking is process-focused, highlighting the importance of generalisable steps, methods, and techniques.

4 Conclusion

The theoretical review shows that the three thinking paradigms are intertwined. It is not possible to design without applying systems thinking and computational thinking. Reviewing some of the notable definitions and descriptions of design reveals how systems thinking and computational thinking are integral to design. Further research and investigation are needed to assess how the three thinking paradigms manifest in a problem context. What is the flow between the three thinking paradigms? How much of that can streamlined and supported with explicit tools and guidelines to support design thinking, systems thinking, and computational thinking?

References

Arnold, R.D., Wade, J.P.: A definition of systems thinking: a systems approach. Procedia Comput. Sci. **4**, 669–678 (2015)

Brown, T.: Design thinking. Harv. Bus. Rev. **86**(6), 84–92 (2008)

Cross, N.: Design thinking: understanding how designers think and work. Berg (2011)

Cross, N., Christiaans, H., Dorst, K.: Analysing design activity. Mater. Des. **16**(2), 122–123 (1996)

Dorst, K., Cross, N.: Creativity in design process: co-evolution of problem-solution. Des. Stud. **22**, 425–437 (2001)

Dorst, K.: The core of 'design thinking' and its application. Des. Stud. **32**(6), 521–532 (2011)

Dorst, K.: Frame innovation: create new thinking by design. MIT press (2015)

Forrester, J.W.: Lessons from system dynamics modelling. Syst. Dyn. Rev. **3**(2), 136–149 (1987)

Gero, J.S.: Design prototypes: a knowledge representation schema for design. AI Mag. **11**(4), 26 (1990)

Granstrand, O., Holgersson, M.: Innovation ecosystems: a conceptual review and a new definition, Technovation 90-91, 102098, ISSN 0166-4972 (2020)

Holling, C.S.: Resilience and stability of ecological systems. Annu. Rev. Ecol. Syst. **4**(1), 1–23 (1973)

Hubka, V. and Eder, W.E. (1988) Theory of Technical Systems, A Total Concept Theory for Engineering Design, Springer Berlin, Heidelberg

Kelly, N., Gero, J.: Design thinking and computational thinking: a dual process model for addressing design problems. Design Science **7**, E8 (2021). https://doi.org/10.1017/dsj.2021.7

Lawson, B.: How designers think, fourth edition: the design process demystified. Architectural Press (2006)
Meadows, D.H.: Thinking in systems. Chelsea Green Publishing (2015)
Moore, J.F.: The death of competition: leadership and strategy in the age of business ecosystems. Wiley Harper Business, New York, NY (1996)
Poon, J., Maher, M.L.: Co-evolution and emergence in design. Artif. Intell. Eng. **11**, 319–327 (1997)
Schon, D.: Designing as a reflective conversation with the materials of a design situation. Knowl.-Based Syst. **5**, 3–14 (1992)
Shute, V.J., Sun, C., Asbell-Clarke, J.: Demystifying computational thinking. Educ. Res. Rev. **22**, 142–158 (2017)
Simon, H.: The sciences of the artificial, 1st edn. MIT Press, Cambridge, Mass (1969)
Simon, H.: Models of bounded rationality, vols. 1 and 2. MIT Press (1982)
Singh, V.: The interaction between design thinking, systems thinking, and computational thinking, In: Chakrabarti, S., Acharya (eds.) The Future of Design Education: Proceedings of InFuSED23, Springer, India (2025)
Wing, J.M.: Computational thinking. Commun. ACM **49**(3), 33–35 (2006)
Wing, J.M.: Computational thinking and thinking about computing. Philosoph. Trans. Royal Soc. A: Math. Phys. Eng. Sci. **366**(1881), 3717–3725 (2008)

Implications of the Co-evolution of Design Research and Design Practice: Strategies for Design Education

Suman Devadula(✉)

Dr.Vishwanath Karad MIT World Peace University, Pune, India
suman.devadula@mitwpu.edu.in

Abstract. The co-evolution of problems and solutions in creative design is understood, theorized and is made part of design education. The co-evolution of design practice and design research is significant and its implications for design education need to be understood. This article explores this co-evolution and examines its implications for design education. By analyzing the historical context of design education briefly and the imperatives like UN sustainable development goals, this paper lays out the necessity to consider more cross-disciplinary approaches for curricular design in design education aspects as one part and also lays out different roles of individual and institutional actors in design education influenced by the co-evolution of design research and design practice. Further, this article proposes strategies for incorporating this co-evolution into design education curricula to better prepare students for complexities of contemporary real-world challenges.

1 Introduction

It is known that knowledge developed even before being produced or discovered through the scientific method alone. While natural arguments about the informal existence of scientific method earlier to its formal establishment, are possible these are inseparable from biased conservative views that can be likened to efforts to prove hypothesis or those efforts that look for evidence in support of the hypothesis foremost. Design existed before science. Design existed even before being industrial and was in the realm of explicit and implicit knowledge of handicrafts. The skill and knowhow in handicrafts, in a funds-striven pre-independent India for e.g., was useful in making early equipment for observations in scientific experiments (Basu 2017). Design Research refers to the systematic inquiry aimed at understanding and improving design processes and outcomes (Archer 1995). It encompasses various methodologies, including empirical studies, theoretical analyses, and experimental approaches, to generate knowledge that can inform practice. Design practice, wherever mentioned here on in this article, unless explicitly stated otherwise, has to be understood to stand for that which is informed from the crafts, that which is informed by necessities of the industrialized professions and also that which is informed by personalization being made possible through contemporary data-driven, rapid and evolving technologies. Design practice involves the application of design principles and methods to create solutions for specific problems,

K. Gericke et al. (Eds.): DSC 2022/2023, PTT, pp. 223–231, 2026.
https://doi.org/10.1007/978-3-032-16278-6_14

with or without the explicit mention of design thinking in real-world contexts, producing products, services, systems, and experiences (Lawson 2006). Design practice has not waited for outcomes for design research and wherever these outcomes of design research are in line with the requirements of design practices, their uptake has been encouraging (Chakrabarti and Lindemann 2016).

Design as a discipline has always been at the intersection of art, science, and technology. The continuous evolution of design practice—driven by changing societal needs, technological advancements, and cultural shifts—has been paralleled by developments in design research (Cross 2001). Design research provides a theoretical foundation and methodological tools that inform and critique design practice (Frayling 1993). The co-evolution of design research and design practice has profound implications for design education, which must adapt to prepare future designers effectively, from the perspective of improving both, i.e. the formal body of knowledge of design by research and of continuing to push the boundaries of design practice.

The purpose of this chapter is to explore how the intertwined evolution of design research and design practice implicates design education. Historical development of design education is briefly reviewed, literature on co-evolution of design research and design practice is analyzed in context, and the implications for design education are summarized before strategizing curricular aspects to accommodate co-evolutionary dynamics of design research and design practice.

2 Background

In the early 20th century, design was considered primarily practice-oriented, with craftsmanship and aesthetics at the forefront. The Bauhaus movement, for example, emphasized the integration of art and industry but did not heavily focus on research (Wick 2000) or even on the user-centered philosophy in its contemporized forms now (Meyer and Norman 2020). Post-World War II saw a shift towards more systematic approaches, with the rise of design methods and the recognition of design as a problem-solving activity (Jones 1992). Herbert Simon and Christopher Alexander contributed to the foundation of design science, advocating for a more research-oriented approach (Simon 1969; Alexander 1964).

In recent decades, the emergence of user-centered design, sustainability, and digital technologies has further blurred the lines between research and practice (Norman 1988; Papanek 1985). The relationship between design research and practice has evolved over time. Designers are now expected to be researchers, strategists, and innovators, integrating insights from multiple disciplines. Design education driven by traditional Bauhaus and Ulm approaches, including their contemporized versions now, are strong in integrating the professional practices and societal considerations respectively. These approaches however have to change gaining from design re-search, design practice and their co-evolution. Co-evolution refers to the reciprocal and iterative influence between two or more entities evolving together (Dorst and Cross 2001). In the context of design, the co-evolution of research and practice means that advancements in one domain stimulate developments in the other. This dynamic relationship fosters innovation and enhances the overall quality and relevance of design outcomes.

From the perspective of design education, the necessity to prepare designers for the future requires content that will be relevant in the near future too. Preparing students at the undergraduate level in this way is important and more so for postgraduate students of design. However, when design research and design practice work in silos, content for design education is ill-informed and the concern for educating designers for the future risks getting lost in one-upmanship between practice and research. As the problem and solution co-evolve in the design process (Wiltsching, et. al. 2013), the concern for educating designers for the future is real and should be informed by a healthy co-evolution between design research and design practice.

3 Co-evolution of Design Research and Design Practice

Design research informs practice by providing evidence-based insights, theories, and frameworks that guide designers (Blessing and Chakrabarti 2009). For instance, research on human-computer interaction (HCI) has profoundly influenced user inter-face design practices (Shneiderman and Plaisant 2010). Conversely, challenges encountered in practice often lead to new research questions, driving the development of novel methodologies and theories (Zimmerman et al. 2007). Below are examples of this co-evolution:

- Sustainable Design: The growing awareness of environmental issues has led to research on sustainable materials and processes (McDonough and Braungart 2002). Designers apply these findings in practice, creating eco-friendly products, which in turn raise new research questions about sustainability.
- Digital Fabrication: Advances in technology, such as 3D printing, emerged from research and have been adopted in practice, transforming prototyping and manufacturing processes (Gibson et al. 2015). Practical applications reveal limitations and opportunities, prompting further research. For e.g. the contribution of 3D printing in innovation allowing rapid prototyping as well as for near-net shape or net-zero manufacturing.
- Design Thinking: Originally a practice-oriented approach, design thinking has been the subject of extensive research, leading to refined methodologies that are taught and applied in various contexts beyond traditional design fields (Brown 2008; Cross 2001).

The boundaries between design research and practice are increasingly permeable. Designers are expected to engage in research activities, and researchers often collaborate with practitioners to test and apply their theories (Sanders and Stappers 2008). This integration has led to more holistic and impactful design solutions but also presents challenges in terms of education and professional development as described further in this article.

4 Implications for Design Education

4.1 Designing Design Education Curriculum

Curricula need to reflect the intertwined nature of research and practice. Traditional models that separate theoretical courses from studio work may no longer be sufficient. Instead, a blended approach that combines theory, design studio methodology, and hands-on projects bringing in real-time constraints within design educational purview can provide a more cohesive learning experience (Davis 2008). Design education requires to plan beyond inter-disciplinarity into trans-disciplinarity involving stakeholders for being purposive and rewarding too.

Emphasizing the co-evolutionary dynamics fosters critical thinking. Students learn to question assumptions, evaluate evidence, and adapt their design processes based on research insights (Nelson and Stolterman 2012). This skill is essential for navigating the rapidly changing landscape of design challenges, particularly when systemic innovation is necessary for addressing grand challenges facing humanity.

4.2 Integrating Research and Practice

The co-evolution of design research and practice necessitates an educational approach that integrates both aspects at different levels of design. Design students must be equipped with the skills to conduct research and apply findings in practical contexts (Friedman 2000). This integration enhances their ability to tackle problems of increasing complexity, involving stakeholders and innovate effectively for real impact. The considerations being different at different levels of design education require a systematic approach to address this challenge. Tables 1 and 2 in this article address this strategic requirement.

4.3 Challenges and Opportunities

Integrating research and practice in education can be resource-intensive and requires faculty with expertise in both areas (Buchanan 2001). Faculty should be encouraged to engage in both research and practice, serving as role models for students (Biggs and Tang 2011). This is challenging for faculty and practitioners who have been conventionally educated and steering as required will require handholding them on pedagogy and its delivery. There may also be resistance to changing established educational models from inured educationists. Embracing the co-evolution can lead to more innovative and relevant educational programs. The criticism that education has not integrated taking up the grand challenges of sustainability aligned with the UN's sustainable development goals (Ossiannilsson 2023) could be addressed. Such integration prepares students to be leaders in the field, capable of driving future co-evolution between design research and design practice (Owen 2007) involved in addressing real problems.

5 Strategies for Design Education

5.1 Project-Based Learning

Project-based learning is very popular with the Bauhaus model of design education so much so that students are allowed to apprentice to an extent of one-fourths of their undergraduate education duration or more. Projects can be structured to address real-world problems, requiring students to investigate, prototype, and iterate solutions informed by research (Blumenfeld et al. 1991). Regulatory bodies for technical higher education in India, regulating design too, have mandated internships (AICTE. 2019) between every academic year and are open through the National Credit Framework for crediting internships and apprenticeships similarly done too (UGC 2023). This has increased the availability of students and the duration of their availability for the industry, although not at a stretch and not after students are industry ready. Design education needs to strategically plan their academic calendar for leveraging regulatory provisions for the benefit of students and increase their acceptance in the industry. This directly means improved exchange between design practice and design academic. However, the in-house requirements of intellectual property (IP) protection for the nature of work for upstream product development activities like design and design research means that design education involvement will be muted to the extend business interests of the design industry (practice) are not compromised.

5.2 Project-Centric Learning

Overcoming the limitations of project-based learning from the design industry perspective can still be good for design students, design industry and hence design re-search. The displaced emphasis on the project rather than on the student temporarily could mean more preference for such education from the industry wherein their time constraints driven by competitive markets can be respected. This learning caters to entrepreneurial aspirations in design students and could mean better uptake of design methods, and hence design research, into design practice. With such design aware, design-preneurs, design education stands to gain virtuously too. Consequently, project centric learning offers strategic leverage in the long-term for design education.

5.3 Collaborative Partnerships for Failure Stories

In spite of the stringent IP mandates of business, continuing to pursue partnerships with design industry, and their related design research teams involved, to share their failures after making already successful designs in the market would provide significant learning opportunities to design students, as they are particularly sensitized to embrace failures whilst challenging the status quo, constructively intervening and learning continuously. This goes beyond the conventional case study mode of learning as it can involve real prototypes and settings planned in good intention but resulted in failures. Establishing partnerships with industry and research institutions in this regard can provide students with valuable opportunities of what does not work and in which contextual setting. Such collaborations raise curiosity and stand to expose design students to the latest developments of design research and their possible practical application (Klein 1990).

5.4 Cross-Disciplinary Mix of Approaches

In education multi-disciplinary, inter-disciplinary and trans-disciplinary approaches could be strategized for the benefit of not only the student but all stakeholders. Consequently, a gradual and a good mix of these approaches are part of the choice architecture for curriculum designers of design education. Table 1 indicates these approaches for the purpose of illustrating the value addition that each approach brings in, whilst gradually progressing in difficulty too.

Table 1. Cross-disciplinary approaches nudge interdisciplinary design education further

	People	Knowledge	Problem solving perspective
Intra-disciplinarity	within discipline only	of and hence within discipline	even if outside discipline, problem will be solved as if solution exists within the discipline
Multi-disciplinarity	People from several and different disciplines work together, each drawing on their disciplinary knowledge	draws on knowledge from different disciplines but staying within their disciplined boundaries	assumes that problem nature can be neatly separated into disciplinary problems. Only such problems and tame problems that need only project managers get solved this way
Interdisciplinarity	experts in more than one discipline being able to see where their knowledge is shared and linked	analogously views discipline from the perspective of another discipline. Analyzes, synthesizes and harmonizes links between disciplines into a coordinated and coherent whole	acknowledges that problem nature cannot always be separated into disciplinary problems and avoids a disciplinary view of the problem foremost. All tame problems and some complex problems get solved this way
Trans-disciplinarity	experts who can take this unified view plan the involvement of all stakeholders to progress on resolving problems involving their disciplinary knowledge in silos, coherently and integratively	In addition to interdisciplinary approach, trans-disciplinarity takes a completely outsider perspective wherein revisiting paradigms and worldviews is also made possible. Embraces naïve realism and integrates knowledge from all stakeholders for the systemic benefit of all in long-term	recognizes the real nature of complex problems and attempts to transcend disciplinary boundaries to integrate knowledges required. This approach goes beyond co-ordinating between the linkages between disciplinary knowledge and acknowledges that real problems can only be resolved temporarily and that when situational contexts change, new problems created therein require further resolutions

With interdisciplinary studies, design students can integrate knowledge from psychology, engineering, business etc., and approach problems more holistically (Groat and Wang 2013). Further, involving all stakeholders makes design education and practice, trans-disciplinary and sustainable too.

6 Discussion

Further to observing the implications, challenges, opportunities and strategizing curricular aspects, it becomes important to systemically discuss the role of individual actors i.e. receivers and providers of design education within the setting of institutions and their actions for individuals and vice versa. The entries in the cells are separated in two blocks/paragraphs so as to be read in this way. Table 2 discusses salient in context considerations that are intended for the mutual benefit of design education from design research and design practice.

Table 2. Endeavor of design students due to co-evolution of design research and practice

	1 Undergraduates (UG)	Post-graduates (PG)	
		2 Masters	3 Doctoral (Ph.D.)
A Design Research (DR)	Participate in DR studies and experiments Inform importance of design having its own body of knowledge (BoK) and initiate into DR within a three-credit course	Participate in DR studies and experiments and involving in planning DR. Involve in DR courses of higher credits Apply outcomes of DR in DP from the perspective of informing DR better in the short-term continuously	Involve all stakeholders appropriately, conduct and publish DR towards formalizing design BoK Develop research design (RD) content, publish it and evolve it continuously for making DR relevant to DP and vice versa. Forecast DP requirements and plan RD for appropriate DR pro-actively. Apply outcomes of DR in DP from the perspective of informing DR better in the long-term continuously. Champion networking of researchers in this context
B Design Practice (DP)	Provide apprenticeship opportunities and inform curriculum to requirements of DP	Apply outcomes of DR in DP for informing DP better and convey feedback to DR, including providing unsolved pragmatic issues of DP	Apply outcomes of DR in DP to resolve issues. Publish this as a case study. Ascertain the roots of origins of issues in DP and address it immediately by DR and publish such DR extensively to prevent such issues in future Invite doctoral students in forecasting studies of design industry and ascertaining DP requirements. Participate in DR towards improving the relation between DP and DR synergistically

(*continued*)

Table 2. (*continued*)

	1 Undergraduates (UG)	Post-graduates (PG) 2 Masters	3 Doctoral (Ph.D.)
C Design Education (DE)	Balance apprenticeship and initiate into Design Research to improve DP	Present case studies of the utility of DR and DP and vice versa for UG students. Assist in DR sessions in context for UG students Involve in planning DE at UG level and represent PG level concerns. Participate in DR involving DR for DP and DE	Educate UG and PG students on RD approaches for DR. Assist in DR sessions alongside faculty in context for PG and UG students Involve in planning DE at UG and PG level. Develop DR content for UG and PG level. Participate in DR towards improving relation between DP and DR synergistically for the benefit of DE

Note: Row heads are Institutional actors and Column heads are Individual actors

7 Summary

The co-evolution of design research and design practice has significant implications for design education. By recognizing and embracing this dynamic relationship, educational institutions can prepare students to be adaptable, capable and innovative in addressing complex real challenges by design. Integrating research and practice within curricula enhances critical thinking, fosters innovation, and aligns education with the evolving demands of the design profession.

Educational strategies such as project-based learning, project-centric learning, collaborative partnerships, cross-disciplinary mix of approaches, with an emphasis on trans-disciplinarity can operationalize the integration of research and practice for addressing real problems of relevance and educating future-ready designers for it. Overcoming challenges related to resource allocation and curricular change is essential for leveraging the opportunities presented by the co-evolution of design research and practice and this is discussed across related aspects laid out in Tables 1 and 2. Design education must evolve in tandem with design research and design practice to remain relevant and effective. By doing so, it will not only equip students with the necessary skills and knowledge but also contribute to the advancement of humanity in sustainable ways pro-actively.

References

Alexander, C.: Notes on the synthesis of form. Harvard University Press, Cambridge, MA (1964)

AICTE. Internship policy, guidelines and procedures (2019). https://www.aicte-india.org/sites/default/files/Aicte%20Internship%20Policy-%2002.04.2019.pdf. Accessed 30 Apr. 2025

Archer, 1995.Archer, B.: The nature of research. Co-Design **2**(11), 6–13 (1995)

Basu, S.: Surface Enhanced Raman Scattering (SERS) Substrates and Probes 2017. https://doi.org/10.13140/RG.2.2.23872.02561. Accessed 30 Apr. 2025

Biggs, J., Tang, C.: Teaching for quality learning at university, 4th edn. McGraw-Hill Education, Maidenhead (2011)

Blessing, L.T.M., Chakrabarti, A.: DRM, a design research methodology. Springer, London (2009)
Blumenfeld, P.C., Soloway, E., Marx, R.W., Krajcik, J.S., Guzdial, M., Palincsar, A.: Motivating project-based learning: sustaining the doing, supporting the learning. Educ. Psychol. **26**(3–4), 369–398 (1991)
Brown, T.: Design thinking. Harv. Bus. Rev. **86**(6), 84–92 (2008)
Buchanan, R.: Design research and the new learning. Des. Issues **17**(4), 3–23 (2001)
Chakrabarti, A., Lindemann, U.: Impact of design research on industrial practice: tools, technology, and training (2016). https://doi.org/10.1007/978-3-319-19449-3. Accessed 30 Apr. 2025
Cross, N.: Designerly ways of knowing: design discipline versus design science. Des. Issues **17**(3), 49–55 (2001)
Davis, M.: Why do we need doctoral study in design? Int. J. Des. **2**(3), 71–79 (2008)
Dorst, K., Cross, N.: Creativity in the design process: co-evolution of problem–solution. Des. Stud. **22**(5), 425–437 (2001)
Frayling, C.: Research in art and design. Royal College Art Res. Papers **1**(1), 1–5 (1993)
Friedman, K.: Creating design knowledge: from research into practice. In: IDATER 2000 Conference. Loughborough University (2000)
Gibson, I., Rosen, D.W., Stucker, B.: Additive manufacturing technologies: 3D printing, rapid prototyping, and direct digital manufacturing, 2nd edn. Springer, New York (2015)
Groat, L., Wang, D.: Architectural research methods, 2nd edn. Wiley, Hoboken, NJ (2013)
Jones, J.C.: Design methods, 2nd edn. Wiley, New York (1992)
Klein, J.T.: Interdisciplinarity: history, theory, & practice. Wayne State University Press, Detroit (1990)
Lawson, B.: How designers think: the design process demystified, 4th edn. Routledge, London (2006)
McDonough, W., Braungart, M.: Cradle to cradle: remaking the way we make things. North Point Press, New York (2002)
Meyer, M., Norman, D.: Changing design education for the 21st century. She Ji: J. Des. Econ. Innov. **6**, 13–49 (2020). https://doi.org/10.1016/j.sheji.2019.12.002
Nelson, H., Stolterman, E.: The design way: intentional change in an unpredictable world, 2nd edn. MIT Press, Cambridge, MA (2012)
Norman, D.A.: The design of everyday things. Basic Books, New York (1988)
Ossiannilsson, E.: Open educational resources (OER) and some of the United Nations sustainable development goals. Int. J. Inform. Learn. Technol. **40** (2023). https://doi.org/10.1108/IJILT-01-2023-0002
Owen, C.L.: Design thinking: notes on its nature and use. Des. Res. Quar. **2**(1), 16–27 (2007)
Papanek, V.: Design for the real world: human ecology and social change, 2nd edn. Thames and Hudson, London (1985)
Sanders, E.B.-N., Stappers, P.J.: Co-creation and the new landscapes of design. CoDesign **4**(1), 5–18 (2008)
Shneiderman, B., Plaisant, C.: Designing the user interface: strategies for effective human-computer interaction, 5th edn. Addison-Wesley, Boston (2010)
Simon, H.A.: The sciences of the artificial. MIT Press, Cambridge, MA (1969)
UGC. National Credit Framework (2023). https://www.ugc.gov.in/pdfnews/9028476_Report-of-National-Credit-Framework.pdf. Accessed 30 Apr. 2025
Wiltschnig, S., Christensen, B., Ball, L.: Collaborative problem–solution co-evolution in creative design. Des. Stud. **34**, 515–542 (2013). https://doi.org/10.1016/j.destud.2013.01.002
Wick, R.K.: Teaching at the Bauhaus. Ostfildern-Ruit: Hatje Cantz (2000)
Zimmerman, J., Forlizzi, J., Evenson, S.: Research through design as a method for interaction design research in HCI. In: Proceedings of the SIGCHI Conference on Human Factors in Computing Systems, pp. 493–502. ACM, San Jose, CA (2007)

Sustainable Design in Higher Education

Daniela Kattwinkel[1(✉)] and Beate Bender[2]

[1] Westfälische Hochschule, Bocholt, Germany
daniela.kattwinkel@w-hs.de
[2] Technische Universität Berlin, Berlin, Germany
beate.bender@tu-berlin.de

Abstract. Reducing global greenhouse gas emissions is a critical goal of our time. Despite a temporary decline during the COVID-19 pandemic, CO_2 emissions are projected to increase again, potentially reaching 42.8 gigatonnes by 2050. Germany aims to reduce its greenhouse gas emissions by 80–95% by 2050, which will require sustainable practices across all industries as they contribute significantly to environmental issues through their products and processes. Engineers play a crucial role in this transition as they are responsible for developing new products, services and product-service systems. Decisions made during the product development process have a significant impact on the environmental impacts of products throughout their life cycle. It is therefore essential that engineers are equipped with the knowledge and skills required for sustainable product development. This chapter examines the integration of sustainability into the engineering education at German universities. The analysis compares data from an initial study conducted in 2018 and a current study conducted in 2023, and examines changes in degree programmes and courses related to sustainable product development at 118 German higher education institutions. By reviewing the current curricula, it aims to understand how universities have adapted to the growing need for sustainability in engineering education and the extent to which current programmes address this crucial topic.

1 Introduction

Reducing global greenhouse gas emissions is one of the central goals of our time (United Nations 2015). The vast majority of these emissions consists of Carbon dioxide (CO_2). Over the past 60 years CO_2 emissions have changed dramatically. Global emissions have increased by more than 270% over this period (Statista Research Department 2022a). Although new data show that emissions fell during the corona-related pandemic years, the level of emissions is expected to recover and rise to 42.8 gigatonnes by 2050 (see Fig. 1) (Statista Research Department 2022b). In Germany, CO_2 emissions have also been reduced in recent years, but are still far from the targets of the Paris Agreement. By signing this document, Germany has committed to reducing its greenhouse gas emissions by 80–95% in the year 2050 compared to the levels of 1990 (Die Bundesregierung 2016). In order to achieve this goal, companies must also act in a more ecological and sustainable manner, as industry as a whole contributes significantly to these emissions

K. Gericke et al. (Eds.): DSC 2022/2023, PTT, pp. 232–247, 2026.
https://doi.org/10.1007/978-3-032-16278-6_15

and environmental problems through its products and processes (Frei 1999). For companies in particular, a sustainable approach offers a new opportunity for technological innovation and economic competitiveness (Die Bundesregierung 2016). This requires the development of new or improved products, services, or product-service systems.

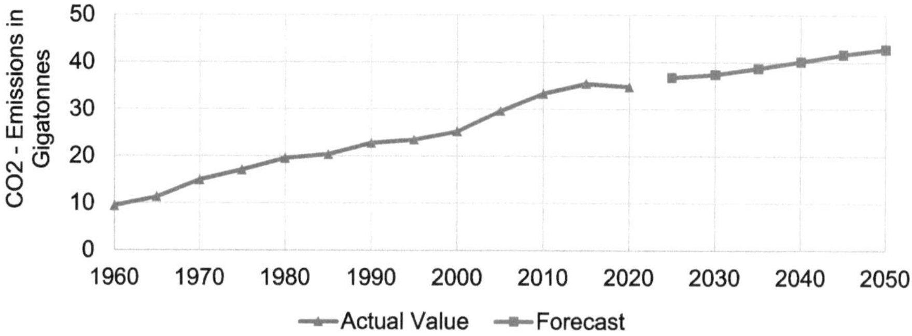

Fig. 1. Real and anticipated worldwide CO_2 emissions from 1960 to 2050 (Statista Research Department 2022a, b)

Many international and EU standards and directives (e.g. the German Energieverbrauchsrelevante-Produkte-Gesetz) have been adopted to support environmentally compatible and sustainable product design in the industry. In addition, a large number of methods and tools have been developed, mainly in academia, but their relevance and application in companies has been limited (Rossi et al. 2016). Even though several authors identified the development of sustainable products as a central requirement to sustain and increase a company's competitive position (Rodrigues et al. 2018). Reasons for the lack of integration of sustainability include, among others, the lack of expertise and experience in the field of sustainable product development and the lack of time to obtain information or training (Vallet et al. 2009). Johansson (2002) defined the expertise and motivation of product development staff as key success factors for the integration of ecodesign and sustainability into the product development process. In this sense, product developers need to pay particular attention to the connection between the company's activities and the resulting environmental impacts (Pigosso 2012). And they need to be trained in the use of appropriate methods and tools (Ritzén 2000). A key starting point for the integration of environmentally compatible and sustainable thinking, methods and tools into business practice is therefore the development of competencies among employees (Danish Environmental Protection Agency et al. 2017). Given the fact that engineers have little time in their working lives to undergo extensive training, it is necessary to promote the development of knowledge and skills in the field of sustainable product development early in the education of future developers. As a result, environmental compatibility and sustainability should be integrated into the training programmes and curricula of future engineers (Kunrath and Ramanujan 2021). The urgency of this issue is also reflected in a recent decision of the European Commission and the German Federal Council (Bundesrat) of 8 April 2022, according to which the teaching of environmental sustainability at universities and throughout the education system is to be integrated in the future in order to respond to the environmental crisis (Bundesrat 2022;

Europäische Kommission 2022). Thus, the education of future product developers and engineers also faces new challenges. Therefore, new educational concepts, attractive for students as well as for the industry have to be developed (Albers et al. 2012). Allenby (2009) even stated that "the combination of [...] unpredictability, complexity, and accelerating change, the interactions of emerging technologies with younger subpopulations, and the demands, however inchoate, of sustainability present a rapidly approaching and inadequately appreciated crisis in education". In the past, however, despite this importance, sustainable product development has hardly been included in the higher education programmes of young engineers and prospective product developers in Germany. This situation is illustrated by an exploratory study conducted prior to this work (Kattwinkel et al. 2018), in which the curricula, syllabi and module handbooks of 118 German universities with a technical background were analysed regarding environmental and sustainable content. In this chapter, we will follow up, if today's courses and degree programs have changed regarding the required sustainability focus by answering the following research questions:

1. How have the degree programmes and courses at German universities changed since the first exploratory study in 2018 with regard to the consideration of sustainability?
2. Which specific sustainability related topics are addressed in current degree programmes and courses?

Following the Design Research Methodology by Blessing and Chakrabarti (2009) in the Research Clarification stage, the role of product development for the reduction of environmental impacts is outlined and required terms are defined (Sect. 2). In the Descriptive Study I, a comprehensive literature and web-based research is conducted to identify courses and degree programs relating to sustainable product development within higher engineering education (Sect. 3.1). In the Prescriptive Study, the courses are evaluated regarding their specific content, and the results are discussed (Sect. 3.2). Finally, the central findings are summarised and possibilities for further research are pointed out (Sect. 4).

2 Research Clarification

2.1 The Role of Product Development

To reduce the global CO_2 emission, sustainable new or improved products, services, or product-service systems have to be generated. These are developed by engineers. As part of the product development process, engineers determine the function, operating principle, form and materials, and thus largely determine the technical, economic and environmental properties of the product (Oberender 2006). Every decision made influences the environmental impact of a product, because every process in the life cycle of a product creates specific environmental impacts (Birkhofer et al. 2018; Oberender 2006). For example, the materials and bond types used determine the manufacturing and assembly processes and influence the disassembly and disposal processes. As a result, material combinations such as adhesive bonds, for example, can make the repair and disassembly of a product considerably more difficult (Fraunhofer-Institut für Fertigungstechnik und Angewandte Materialforschung 2020). Therefore, in addition to technical and design-specific knowledge and skills, the development of sustainable products also requires extensive knowledge of the product life cycle phases, their processes, and their interactions. The processes and their environmental impacts throughout the entire product life cycle are largely determined during the product development and therefore long before a product is manufactured (Oberender 2006). It is estimated that a significant proportion, ranging from 60% to 80%, of a product's environmental impact is determined during the initial phases of the product development (Pigosso 2012). The influence of the product development on the environmental impacts of products is shown schematically in Fig. 2. Future information on the environmental impact of a product is increasingly difficult to predict, e.g. disposal scenarios that may not materialise for several decades.

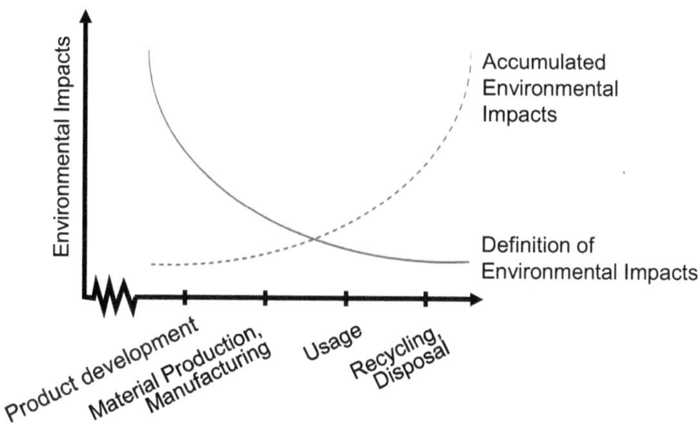

Fig. 2. Definition and origin of environmental impacts (Dannheim 1999)

Developers are faced with the challenge of anticipating and addressing various processes and their effects throughout all stages of the product life cycle. They must select, design, and optimize processes in advance to ensure the development of sustainable products. This presents a significant challenge for developers, as they need to balance multiple and sometimes conflicting criteria while adopting a pragmatic approach to meet time and cost objectives (Tischner and Moser 2015).

2.2 Sustainable Product Development and Ecodesign

In public discourse, "eco", "green" or "environmentally compatible" are often used as synonyms for "sustainable" (Birkhofer et al. 2018). However, *sustainability* encompasses a much larger development of society as a whole that allows current generations to meet their needs without depriving future generations of these opportunities (United Nations 2015). Accordingly, sustainable product development seeks to take into account the needs of future generations and considers economic, ecological and social aspects (Charter and Tischner 2001). A holistically sustainable development process requires the integral and equal consideration of the dimensions of economy, ecology and society as well as their interactions (Coric 2016). *Design for Sustainability (DfS)* or *sustainable product development* thus, generates solutions that deliver meaningful social benefits, add value for both suppliers and consumers, and prevent ecological harm or potentially contribute positively to the natural environment (Charter and Tischner 2001). Various approaches, such as *Design for Environment (DfE), life cycle design, green design, eco-effective product development, Ecodesign,* and *environmentally compatible product development,* share a common objective—reducing the environmental impacts of products, with a focus on ecological and economic aspects (McAloone and Pigosso 2021; Zhao 2013). These methodologies typically involve systematic procedures in which ecological considerations are integrated at the early stages of product planning, development, and design. The aim is to diminish the environmental impact of a product across its entire life cycle without compromising other essential criteria (Johansson 2002). In other approaches, such as the *circular economy,* the aim is to use resources for as long as possible in order to minimise the material and energy requirements as well as the waste and emissions. By means of reuse and recovery, recycling and resource-efficient product design (or Ecodesign), these goals can be achieved (Neligan et al. 2023).

Sustainability has also become increasingly important for governments. In 2015 the 193 member states of the United Nations (UN) adopted the 2030 Agenda for Sustainable Development (United Nations 2015). This agenda and its 17 Sustainable Development Goals (SDGs) form the basis of the German government's sustainability policy. The further developed German Sustainability Strategy (Deutsche Nachhaltigkeitsstrategie) includes the measures with which the Federal Government has responded to the COVID-19 pandemic and with which the conditions for a sustainable future are being strengthened (Die Bundesregierung 2021). This goes hand in hand with the development of specific key competences as a prerequisite for participation in the learning and communication processes for sustainable development. Education plays a central role in the realisation of this transformation, as it not only prepares individuals to tackle global sustainability problems, but also promotes the collective process of sustainable development. The aim of Education for Sustainable Development is to raise awareness of sustainability-related challenges and to promote the acquisition of key competences in order to enable people to actively and responsibly participate in and contribute to sustainable development and responsible participation in shaping sustainable development (Frye 2022).

2.3 Sustainability in the Industry

In 2018, a large-scale study by the Institute for Employment Research (Institut für Arbeitsmarkt- und Berufsforschung), a research institute of the Federal Employment Agency (Bundesagentur für Arbeit) that surveys around 16,000 companies in Germany every year, examined environmental sustainability in companies (Bellmann et al. 2019). The results of the survey show that the importance of environmental sustainability varies according to the number of employees in a company. In companies with one to four employees, more than 45% of all employees state that environmental sustainability is very important or extremely important for their business activities. For companies with more than five employees, the figure is between 50 and 60%, and for companies with more than 200 employees, the figure is above 60%. In companies with more than 5,000 employees, more than 75% of respondents said sustainability was very important (55%)

and important (23%). So, it was generally found that environmental sustainability is more important in larger companies than in smaller ones.

Another more current empirical study is based on a company survey conducted in summer 2022 by the Institute of the German Economy, in which 1,200 companies were asked about the role of products and/or services that strengthen circularity and the underlying motivations and approaches (Neligan et al. 2023). Among other things, the companies were asked what role environmentally friendly product features play in their company's product range (see Fig. 3). Around 60% of manufacturing companies with and without a service offering consider low resource consumption or high resource efficiency in relation to their own use of resource during the production to be fundamentally important - including almost 30% of product and service providers who even stated that it was "very important". For the transition to more sustainable products, it is essential to take their entire product life cycle into account and to reduce the environmental impacts along this cycle. More than half of the manufacturing companies consider low environmental impacts to be very important or important, and over 60% of product and service providers do so. The authors were able to show in an earlier company survey that companies aim to increase resource efficiency primarily to reduce manufacturing costs, avoid waste and protect the environment by conserving resources (Neligan et al. 2023).

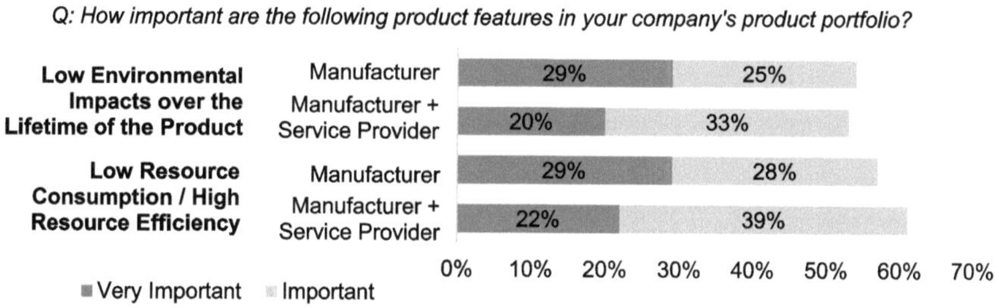

Fig. 3. Excerpt from a survey on the importance of a reduction of the resource consumption and the environmental impacts (Neligan et al. 2023)

Overall, the two studies were able to show, that companies are concerned with implementing more sustainability within the company itself, as well as through their products and services. They aim to consume fewer resources and thus reduce environmental impacts. In the past, this was primarily done reactively, for example, in response to laws, customer demands, or to reduce raw material shortages and dependencies from suppliers. Now, companies also understand the potential of a proactive sustainable product and service design to create new market opportunities, secure or enhance competitiveness, and leverage new possibilities arising from digitisation (Neligan et al. 2023). This means, companies will need graduates especially from the field of engineering that can contribute to realising their sustainability goals in the future.

3 Sustainable Design Courses at Higher Education Institutions

3.1 Material and Method

The curricula of German universities were examined manually with regard to courses that emphasize sustainable topics in an engineering context. A comprehensive list of 120 institutions with a technical focus that offer university degrees in engineering (B.Sc., M.Sc., B.Eng., M.Eng.) was extracted from all German universities and universities of applied science (Ger: Fachhochschulen) (Bohlken 2017a, b). The list has been thoroughly reviewed and verified for accuracy, including the correction of any duplicate names, renames, and name changes (see Fig. 4). It serves as a basis for analysing the range of courses and degree programs and includes 118 universities after the revision.

1. AKAD Hochschule
2. bbw Hochschule Berlin
3. Berliner Hochschule für Technik
4. Brandenburgische Technische Universität Cottbus-Senftenberg
5. DIPLOMA Hochschule Studienzentrum Berlin
6. Duale Hochschule Baden-Württemberg
7. Ernst-Abbe-Hochschule Jena
8. Europäische Studienakademie Kälte-Klima-Lüftung ESaK Maintal
9. Fachhochschule Münster
10. Fachhochschule Schmalkalden
11. Fachhochschule Südwestfalen, Iserlohn
12. Fachhochschule Trier – Umwelt-Campus Birkenfeld
13. FH Aachen
14. FH Dortmund
15. FH Kiel
16. Frankfurt University of Applied Sciences
17. Friedrich-Alexander-Universität Erlangen-Nürnberg
18. HAW Hamburg
19. Helmut-Schmidt-Universität – Universität der Bundeswehr Hamburg
20. Hochschule Aalen
21. Hochschule Ansbach
22. Hochschule Bielefeld
23. Hochschule Bochum
24. Hochschule Bonn-Rhein-Sieg
25. Hochschule Bremen
26. Hochschule Bremerhaven
27. Hochschule Coburg
28. Hochschule Darmstadt
29. Hochschule Düsseldorf
30. Hochschule Emden/Leer
31. Hochschule Esslingen
32. Hochschule Flensburg
33. Hochschule für angewandte Wissenschaft und Kunst Hildesheim/Holzminden/Göttingen
34. Hochschule für angewandte Wissenschaften München
35. Hochschule für Technik und Wirtschaft Berlin
36. Hochschule für Technik und Wirtschaft des Saarlandes
37. Hochschule für Technik und Wirtschaft Dresden
38. Hochschule für Technik, Wirtschaft und Kultur Leipzig
39. Hochschule Hamm-Lippstadt
40. Hochschule Hannover
41. Hochschule Heilbronn
42. Hochschule Hof
43. Hochschule Kaiserslautern
44. Hochschule Karlsruhe
45. Hochschule Kempten
46. Hochschule Koblenz
47. Hochschule Konstanz Technik, Wirtschaft und Gestaltung
48. Hochschule Landshut
49. Hochschule Mannheim
50. Hochschule Mittweida
51. Hochschule Niederrhein
52. Hochschule Nordhausen
53. Hochschule Offenburg
54. Hochschule Osnabrück
55. Hochschule Pforzheim
56. Hochschule Ravensburg-Weingarten
57. Hochschule Reutlingen
58. Hochschule Rhein Waal
59. Hochschule RheinMain
60. Hochschule Ruhr West
61. Hochschule Stralsund
62. Hochschule Trier
63. Hochschule Wismar
64. Hochschule Zittau/Görlitz
65. HS Anhalt
66. HS Magdeburg-Stendal
67. HS Merseburg
68. Jade Hochschule
69. Karlsruher Institut für Technologie
70. Leibniz Universität Hannover
71. Ostbayerische Technische Hochschule Amberg-Weiden
72. Ostbayerische Technische Hochschule Regensburg
73. Ostfalia Hochschule für angewandte Wissenschaften
74. Otto-von-Guericke-Universität Magdeburg
75. Private Hochschule für Wirtschaft und Technik
76. Rheinische Fachhochschule Köln
77. Ruhr-Universität Bochum
78. RWTH Aachen
79. Technische Fachhochschule Georg Agricola
80. Technische Hochschule Aschaffenburg
81. Technische Hochschule Augsburg
82. Technische Hochschule Ingolstadt
83. Technische Hochschule Lübeck
84. Technische Hochschule Ostwestfalen-Lippe
85. Technische Hochschule Rosenheim
86. Technische Hochschule Ulm
87. Technische Hochschule Wildau
88. Technische Hochschule Würzburg-Schweinfurt
89. Technische Universität Clausthal
90. Technische Universität Darmstadt
91. Technische Universität Hamburg
92. Technische Universität München
93. TH Bingen
94. TH Brandenburg
95. TH Deggendorf
96. TH Köln
97. TH Mittelhessen
98. TH Nürnberg Georg Simon Ohm
99. TU Bergakademie Freiberg
100. TU Berlin
101. TU Braunschweig
102. TU Chemnitz
103. TU Dortmund
104. TU Dresden
105. TU Ilmenau
106. TU Kaiserslautern
107. Universität Bayreuth
108. Universität Bremen
109. Universität der Bundeswehr München
110. Universität des Saarlandes
111. Universität Duisburg-Essen
112. Universität Kassel
113. Universität Paderborn
114. Universität Rostock
115. Universität Siegen
116. Universität Stuttgart
117. Westfälische Hochschule Gelsenkirchen, Bocholt, Recklinghausen
118. Wilhelm Büchner Hochschule

Fig. 4. List of German universities with a technical focus (based on (Bohlken 2017a, b))

The analysis of available courses involves reviewing the websites, curricula, syllabuses, course catalogues, and module handbooks for Bachelor's, Master's, and elective modules. Specific topics are identified through an analysis of the data using predefined codes and keywords. The keywords derive from the definitions of the basic terminology in Sect. 2. Related terms and synonyms, such as 'ecodesign' for ecological product design, as well as English and German terms, such as 'sustainability' and 'Nachhaltigkeit', are also considered. To aid in the analysis of the text documents, search phrases based on Frye (2022) are assigned to keywords to simplify the search, as shown in Table 1. The search was conducted with the search function of the PDF reader.

Table 1. Search phrases and keywords/codes for the analysis (Frye 2022)

Search Phrase	Keywords/Codes	Search Phrase	Keywords/Codes
circular*	Circular, circular economy, …	**ökonom***	Ökonomisch, Ökonomie, …
eco*	Ecodesign (Ökodesign), eco-design, ecology, ecological, economy, …	**produ***	Produkt, Produktentwicklung, product development, product design, Produktlebenszyklus, …
environment*	Environment, environmental, environmentally compatible, …	**recycl***	Recycling, recycle, …
Klima*	Klima, Klimaschutz, Klimawandel, Klimaziele, …	**social***	Social, Corporate social responsibility, …
Kreislauf*	Kreislaufwirtschaft, …	**Sozial***	Sozial, Soziales, …
nachhaltig*	Nachhaltigkeit, nachhaltig, nachhaltigkeitsorientiert, …	**sustainab***	Sustainability, sustainable, sustainable design, …
ökolog*	Ökologisch, Ökologie, …	**Umwelt***	Umwelt, Umweltschutz, umweltschädlich, …

3.2 Results of the Analysis

3.2.1 Sustainable Degree Programmes

Based on the content of the module manuals, the handbooks, and descriptions on the websites of the universities, we examined the courses and the degree programmes regarding their subject matter. The results of the analysis are illustrated in Fig. 5. In 2018, 120 degree programmes could be investigated. In 2023, the degree programmes and courses of 118 higher education institutions were examined.

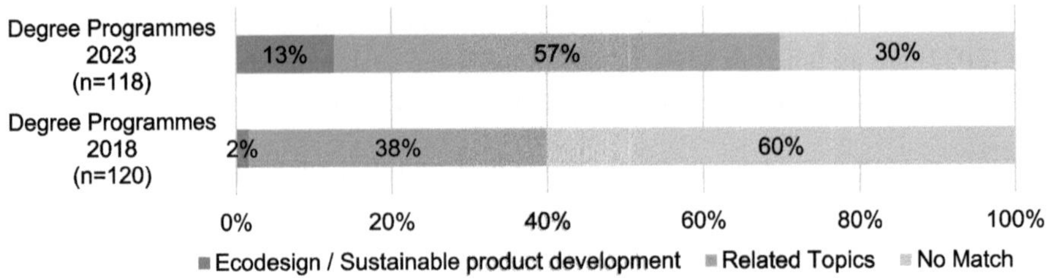

Fig. 5. Analysis of degree programmes in the years 2023 and 2018

In the year 2018 only two of the analysed universities in Germany (2%) offered a degree programme in the field of Ecodesign or sustainable product development or design. Those were *Sustainable Engineering Science* (Hochschule Bonn-Rhein Sieg) and *Resource Efficiency in Mechanical Engineering* (Hochschule Esslingen). 46 universities (38%) offered degree programmes on related topics and the majority of 72 universities (60%) did not offer degree programmes relating to sustainability. In 2023, 15 out of the 118 analysed German universities (13%) offer degree programmes in the field of Ecodesign and sustainable product development. Altogether, 20 bachelor's and master's degree programmes (B.Sc./M.Sc. and B.Eng./M. Eng.) could be identified. Table 2 provides examples of those identified degree programmes, these include the two degree programmes that already existed in 2018. At the Bochum University of Applied Sciences (Hochschule Bochum) it is for example possible to receive a Bachelor's and a Master's degree in *Sustainable Development* and a Master's degree in *Applied Sustainability* The sustainability programmes approach the topic of sustainability from the perspective of natural sciences, economics, engineering and social sciences. Their educational objective is to enable graduates to optimise internal company processes and procedures in the context of sustainability. They are also qualified to develop and analyse products and services meeting sustainability requirements. Sustainability reporting and effective communication with the company's external stakeholders are also key components of the training programme (Hochschule Bochum 2024a, b, c). The Technical University of Braunschweig (Technische Universität Braunschweig) for example, offers the degree programme *Sustainable Engineering of Products and Processes* (B.Sc.), which was designed to meet the changing demands placed on the professional activities of engineers. The aim of this degree programme is to equip future engineers with practical knowledge of evaluation and assessment methods, as well as technical knowledge and engineering skills. This involves the entire life cycle of products and systems, taking into account their economic and social impact. This curriculum enables students to develop sustainable and successful strategies for solutions in areas such as energy supply, process engineering, mobility, product design, and manufacturing (Technische Universiät Braunschweig 2024). The *Sustainable Engineering and Management* (B.Sc.) programme at the Westphalian University of Applied Sciences (Westfälische Hochschule) qualifies students in the field of sustainable value creation. The focus lies on the environmental, economic, and social impact of companies. Students learn to assess sustainability in various areas of the company, including supply chains, production processes and corporate management. The aim of the programme is to enable graduates to initiate sustainable improvements in the company and thus make a positive contribution to changing the future (Westfälische Hochschule 2024).

Table 2. Examples of sustainable degree programmes at technical universities in 2023

University	Degree Programme	Degree
Hochschule Bochum	Sustainable Development	B.Sc./M.Sc
	Applied Sustainability	M.Sc
Hochschule Bonn-Rhein-Sieg	Sustainable Engineering Sciences	B. Eng./M. Eng
Hochschule Esslingen	Resource Efficiency in Mechanical Engineering	B.Sc
Hochschule Zittau/Görlitz	Green Engineering	B. Eng
Technische Universität Darmstadt	Mechanical Engineering – Sustainable Engineering	B.Sc
Technische Universität Braunschweig	Sustainable Engineering of Products and Processes	B.Sc
Westfälische Hochschule	Sustainable Engineering and Management	B.Sc

Overall, the analysis showed that universities of applied sciences in particular offer sustainable engineering degree programmes. Figure 6 on the left illustrates the ratio of sustainable degree programmes offered at three universities and 12 universities of applied sciences. The majority of the degree programmes is completed with a Bachelor's degree (65%) and only about one third (35%) are Master's programmes, as illustrated in Fig. 6 on the right.

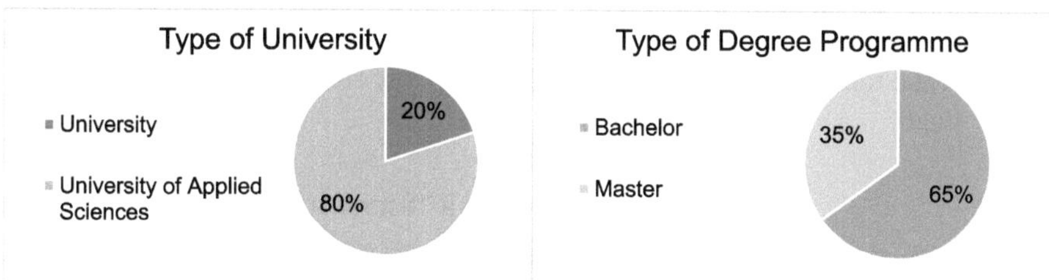

Fig. 6. Comparison of the type of university (left) and the type of degree programmes (right) for the sustainable degree programmes from 2023

In 2023, the majority of 68 German universities (57%) offered engineering or similar degree programmes that focus on topics related to ecodesign, sustainable design, sustainable engineering or green engineering (see Fig. 5). The analysis revealed that a large number of these degree programmes are thematically associated with the subject area "energy". Examples include the degree programmes *Regenerative Energy Systems*

(Technische Universität Berlin) or *Sustainable Energy Systems* (Technische Universität Dortmund). In this context, the results of the analysis also showed that degree programmes, particularly in the field of electrical engineering or electromobility are often offered in combination with topics such as energy efficiency. Examples are *Energy Efficiency and Renewable Energies - Electrical Engineering* (Frankfurt University of Applied Sciences), *Electromobility and Regenerative Energy Technology* (Technische Universität Chemnitz), or *Sustainable Energy Systems and Electromobility* (Technische Universität Braunschweig). However, there are still a total of 35 universities (30%) which do not offer degree programme related to sustainability. However, compared to the 2018 analysis, the number of universities with degree programmes on Ecodesign and sustainable product development has increased from two to 15 institutions and the number of universities with sustainability-related degree programmes has increased from 46 to 68 institutions.

3.2.2 Sustainable Courses

In addition to the degree programmes, in a following detailed analysis we also examined whether the universities offer individual courses on sustainable topics. For this purpose, we reviewed the courses offered by 118 German universities in 2018 and 2023. The results are illustrated in Fig. 7. In Germany, only eleven universities (9%) had courses focusing on Ecodesign or sustainable product development and design in their curricula (Fig. 8). Courses that showed a very high conformity with the topics Ecodesign and sustainable design were for example, *Ergonomics and Ecodesign* (Hochschule Aalen) or *Resource-efficient Product Development* (Technische Universität Clausthal). The majority of 98 German universities (83%) offered engineering courses on related environmental topics, such as *Innovative and Energy efficient Electric Drives for Railway Systems* (Technische Hochschule Brandenburg), *Lean Production* (Hochschule Ansbach) or *Renewable Energies* (Ruhr-Universität Bochum). A total of nine universities did not offer any sustainability related courses in 2018. By 2023, these numbers have changed dramatically. A total of 55 universities (46%) has now included courses on Ecodesign and sustainable product development and 61 universities (52%) offer courses on related environmental topics. Only two universities still have no sustainability related course in their catalogue.

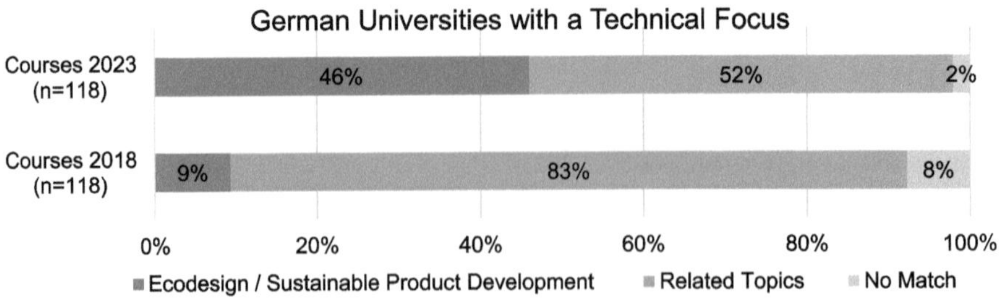

Fig. 7. Analysis of university courses in the years 2023 and 2018

In a more detailed analysis, all identified courses on sustainable topics in 2018 (a total of 356 courses) and 2023 (a total of 595 courses) were grouped in general topics. To do this, the course descriptions in the module handbooks were examined and eleven main themes were identified. The next step was to assign the courses to these themes and calculate the percentage distribution, as shown in Fig. 8. One of the main findings of this study is that significantly more courses now cover Ecodesign and sustainable product development (11%) than in 2018 (3%). In the study from 2023 a total of 123 courses could be identified for this topic. For example, the Master's course *Development Methods for Sustainable Products and Sustainable Product Development - Blue Engineering* (TU Berlin) or *Sustainable Product Development* (Ruhr-Universität Bochum) is a Master's course to promote an understanding of the sustainable design of technical products using real product examples. More technical topics such as Sustainable Energy Technologies (SET) and Lightweight Engineering (LE) are addressed less (SET: 12% and LE: 4%) than in the first study (SET: 21% and LE: 17%). Instead, subordinate topics such as Energy and Energy and Climate Management are now increasingly addressed by courses (26%) compared to 2018 (21%). There are also more courses covering Recycling and Disposal (12%) than in the first study (6%). It is also interesting to note the vast increase in courses on Ethics and Sustainability (13%) compared to the 2018 study (2%). In the topics Alternative Drive Technologies, Environment and Climate, Sustainability Management, Sustainable Manufacturing and Sustainable Materials, there were only subject to slight changes from 2018 to 2023.

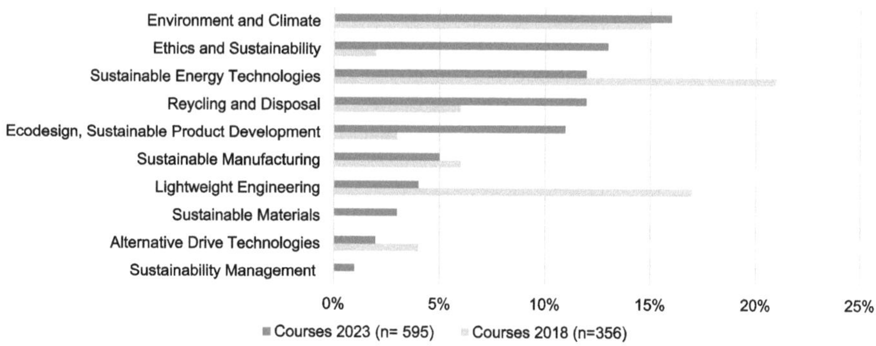

Fig. 8. Overview of the general topics of the analysed courses

Overall, the results of the two studies and their comparison showed an increase in sustainable degree programmes and courses offered at German universities. It can also be noted that the number of sustainability-related courses has increased much faster than the number of degree programmes. This is due to the fact that it is significantly more complex, and time consuming to develop and accredit a whole degree programme compared to offering additional individual courses. The number of sustainable degree programmes at universities of applied sciences has also increased. This might correlate with the facts that universities of applied sciences work more closely with the industry or local companies and are thus able to identify their needs easier. Often, universities of applied sciences, have a lower number of students and staff members, such as the Bochum

University of Applied Sciences, where around 7.500 students are enrolled and 550 staff members work (Hochschule Bochum 2024d). This might facilitate change and with that the development of new degree programmes, compared to universities with almost 40.000 students and 6.300 members of staff, such as the Ruhr-University Bochum (Ruhr-Universität Bochum 2023). The higher number of Bachelor's programmes addressing sustainability may be due to the fact that bachelor's programmes are developed first and then master's programmes. Since the two studies were carried out only five years apart, it is also possible that the master's programmes have not yet been completed.

The results of the two studies conducted may differ from each other as different examination methods were used. In 2018, all module handbooks, course descriptions and syllabi were analysed and examined by people. In 2023, the search function of the PDF reader was used instead, with the search phrases described in Table 1. In particular, this may mean that courses with more creative names or names that contain abbreviations, might not have been identified in the 2018 analysis, but could have bend found by the automated search in 2023, in which the entire text of the module handbooks was automatically analysed (e.g. the abbreviation AFV for Alternatively fuelled vehicle). In addition, this investigation is based on the hypothesis that the content taught at universities will eventually lead to an increased implementation of these topics in the industry. The transfer of design methodologies from research to industry is not part of this research (as many methodologies and methods proposed by researchers to support product developers have not been transferred into practice) and the didactic format of the courses analysed was also neglected. Classical lectures are the leading teaching format in engineering (Kossack and Bender 2024). Although most professors value practical design projects for teaching constructive skills, mostly due to large numbers of students teacher-centred lectures will continue to dominate in the in the future (Albers et al. 2012).

4 Conclusions

In conclusion, industrial, political and social changes demand a stronger integration of sustainable topics into the education of engineers who will develop the products of tomorrow and thus shape the world, in which we will be living in the future. This integration was still in its infancy in 2018. The renewed analysis of the curricula, syllabuses and module handbooks offered by German universities in 2023 showed more and more universities offering degree programmes and courses on sustainability and related topics for engineering students. There has been a particular increase in the number of programmes and courses, not only in the field of Ecodesign and sustainable product development, but also in a wider societal context that has been less addressed in the past. These adaptations are crucial in order to prepare students for current and future challenges and to promote environmentally conscious education. However, there is still a long way to go before Ecodesign and sustainable design become an integral part of general engineering education and not just an "add-on" or a specialisation. In addition to the mere sharing of (technical) knowledge, courses should address topics as complex as sustainability with appropriate teaching (e.g. project-based learning) and assessment formats (e.g. oral presentation) to lead to an increase in skills as well as competences and to avoid a pure memorisation of facts. New, innovative teaching and learning concepts are therefore required to enable a long-lasting learning success and to empower students to apply their sustainable skills and competences in their professional future (see Kattwinkel 2023). One way for universities to facilitate this change is to open up teaching materials and share them with other educational institutions (e.g. via online platforms). To this end, universities and their staff need to increase their willingness to disseminate and share information and to work more closely together (e.g. interdisciplinary educational focus groups). Future analysis should identify, through quantitative (e.g. surveys) or qualitative (e.g. expert interviews) studies, which competences, skills or know-how companies need from engineering graduates in the field of sustainability. This could serve as a basis for the comparison between the demands of the industry and the supply of the university.

References

Albers, A., Denkena, B., Matthiesen, S.: Faszination Konstruktion: Berufsbild und Tätigkeitsfeld im Wandel (arcatech STUDIE) (2012)

Allenby, B.R.: The Challenge of Sustainable Engineering Education. Symposium on Engineering and Liberal Education. Union College. New York (2009)

Bellmann, L., Koch, T.: Ökologische Nachhaltigkeit in deutschen Unternehmen: Empirische Ergebnisse auf Basis des IAB-Betriebspanels 2018, IAB-Forschungsbericht, No. 8/2019, Institut für Arbeitsmarkt- und Berufsforschung (IAB), Nürnberg (2019)

Birkhofer, H., Schulze, K., Zhao, S., Sarnes, J.: Umweltgerechtes Konstruieren. In: Rieg, F. (ed.) Handbuch Konstruktion, 2nd edn., pp. 599–620. Hanser, München (2018)

Blessing, L.T.M., Chakrabarti, A.: DRM, a design research methodology. New York, Springer, Dordrecht (2009)

Bohlken, J.: Hochschulen in Deutschland - Universitäten, eine Initiative des Profiling Institut Düsseldorf (2017a). http://www.hochschulen-deutschland.org/fachhochschulen.html. Accessed 11 June 2017

Bohlken, J.: Hochschulen in Deutschland - Universitäten, eine Initiative des Profiling Institut Düsseldorf (2017b). http://www.hochschulen-deutschland.org/universitaeten.html. Accessed 11 June 2017

Bundesrat. Vorschlag für eine Empfehlung des Rates zum Lernen für ökologische Nachhaltigkeit. Drucksache 13/22 (2022)

Charter, M., Tischner, U. (eds.): Sustainable solutions: developing products and services for the future. Taylor and Francis, London (2001)

Coric, M.: Die Entwicklung nachhaltiger Produkte: Ein ganzheitlicher methodischer Ansatz der Nachhaltigkeitsbilanz, Dissertation, Darmstadt, Technische Universität Darmstadt (2016). https://d-nb.info/1112269754/34. Accessed 4 May 2021

Danish Environmental Protection Agency, Confederation of Danish Industry, IPU and DTU (2017) iMPU Survey results. http://www.ecodesign.dtu.dk/iMPU/Competences. Accessed 5 June 2020

Dannheim, F.: Die Entwicklung umweltgerechter Produkte im Spannungsfeld von Ökologie und Ökonomie: Eine Analyse unter besonderer Berücksichtigung der Nutzungsphase (Zugl.: Darmstadt, Techn. Univ., Diss., 1999), Düsseldorf, VDI-Verl (1999)

Die Bundesregierung (ed). Deutsche Nachhaltigkeitsstrategie - Neuauflage 2016, Berlin (2016). https://www.bundesregierung.de/resource/blob/975292/730844/3d30c6c2875a9a08d364620ab7916af6/deutsche-nachhaltigkeitsstrategie-neuauflage-2016-download-bpa-data.pdf?download=1. Accessed 3 April 2020

Die Bundesregierung (ed). Deutsche Nachhaltigkeitsstrategie Weiterentwicklung 2021, Berlin (2021)

Europäische Kommission. Vorschlag für eine Empfehlung des Rates zum Lernen für ökologische Nachhaltigkeit. Drucksache 13/22 (2022)

Fraunhofer-Institut für Fertigungstechnik und Angewandte Materialforschung: Kreislaufwirtschaft und Klebtechnik: Studie des Fraunhofer-Instituts für Fertigungstechnik und Angewandte Materialforschung IFAM. Fraunhofer Verlag, Stuttgart (2020)

Frei, M.: Öko-effektive Produktentwicklung: Grundlagen - Innovationsprozess - Umsetzung. Gabler, Wiesbaden (1999)

Frye, S.: Förderung nachhaltigkeitsorientierter Kompetenzen im ingenieurwissenschaftlichen Studium, Dissertation, Technische Universität Dortmund (2022)

Hochschule Bochum (ed). Angewandte Nachhaltigkeit: Master of Science (2024a). https://www.hochschule-bochum.de/studium/studienangebote/masterstudiengaenge/angewandte-nachhaltigkeit/. Accessed 31 Jan 2024

Hochschule Bochum (ed). Nachhaltige Entwicklung: Bachelor of Science (2024b). https://www.hochschule-bochum.de/nachhaltige-entwicklung/

Hochschule Bochum (ed). Nachhaltige Entwicklung: Master of Science (2024c). https://www.hochschule-bochum.de/studium/studienangebote/masterstudiengaenge/nachhaltige-entwicklung/. Accessed 31 Jan 2024

Hochschule Bochum (ed). Zahlen und Fakten (2024d). https://www.hochschule-bochum.de/die-bo/hochschule/profil/zahlen-und-fakten/. Accessed 31 Jan 2024

Johansson, G.: Success factors for integration of ecodesign in product development. Environ. Manag. Health **13**(1), 98–107 (2002)

Kattwinkel, D.: Umweltgerechte Produktentwicklung - Systematische Entwicklung, Umsetzung und Evaluation eines Lehrveranstaltungskonzepts in den Ingenieurwissenschaften (Hochschulschrift), München, Dr. Hut (2023)

Kattwinkel, D., Song, Y.-W., Bender, B.: Analysis of Ecodesign and sustainable design in higher education. In: Proceedings of the DESIGN 2018, pp. 2451–2460 (2018)

Kossack, F., Bender, B.: Engineering design education at german universitys: potential for a common basis to create e-learning content. In: Proceedings of the DESIGN 2024 18th International Design Conference (2024)

Kunrath, K., Ramanujan, D.: Fostering sustainable mindsets in engineering education. In: Proceedings of the Design Society, vol. 1, pp. 1597–1606 (2021)

McAloone, T.C., Pigosso, D.C.: Ökodesign: Entwicklung von Produkten mit verbesserter Ökobilanz. In: Bender, B., Gericke, K. (eds.) Pahl/Beitz Konstruktionslehre, pp. 975–1021. Springer, Berlin, Heidelberg (2021)

Neligan, A., Lichtenthäler, S., Schmitz, E.: Produkte und Di enste für eine zirkuläre Wirtschaft: Ergebnisse aus dem IW-Zukunftspanel, IW-Report 16/23 (2023)

Oberender, C.: Die Nutzungsphase und ihre Bedeutung für die Entwicklung umweltgerechter Produkte. VDI Verlag, Düsseldorf (2006)

Pigosso, D.: Ecodesign maturity model: a framework to support companies in the selection and implementation of ecodesign practices. University of Sao Paulo, Sao Carlos (2012)

Ritzén, S.: Integrating environmental aspects into product development: proactive measures, Dissertation, Stockholm, Royal Institute of Technology (2000)

Rodrigues, V., Pigosso, D., Andersen, J., McAloone, T.: Evaluating the potential business benefits of ecodesign implementation: a logic model approach. Sustainability **10**(6), 2011 (2018)

Rossi, M., Germani, M., Zamagni, A.: Review of ecodesign methods and tools. Barriers and strategies for an effective implementation in industrial companies. J. Clean. Product. **129**, 361–373 (2016)

Ruhr-Universität Bochum (ed). Zahlen und Fakten: Steckbrief (2023). https://uni.ruhr-uni-bochum.de/de/zahlen-und-fakten. Accessed 31 Jan 2024

Statista Research Department (2022a). CO2-Emissionen weltweit in den Jahren 1960 bis 2020. https://de.statista.com/statistik/daten/studie/37187/umfrage/der-weltweite-co2-ausstoss-seit-1751/#professional. Accessed 2 May 2022

Statista Research Department. Prognose zu den energiebedingten Kohlendioxid-Emissionen weltweit in den Jahren 2020 bis 2050 (2022b). https://de.statista.com/statistik/daten/studie/28937/umfrage/prognose-zur-kohlendioxid-emission-weltweit-bis-2050/. Accessed 4 May 2022

Technische Universität Braunschweig (ed). Sustainable Engineering of Products and Processes (2024). https://www.tu-braunschweig.de/studienangebot/sustainable-engineering-of-products-and-processes-bachelor. Accessed 31 Jan. 2024

United Nations (ed). TRANSFORMING OUR WORLD: THE 2030 AGENDA FOR SUSTAINABLE DEVELOPMENT: A/RES/70/1, 2015 (2015)

Vallet, F., Millet, D., Eynard, B.: Investigation the use of eco-design guides: presentation of two case studies,. In: International Conference on Engineering Design, ICED'09; 24–27 August 2009 pp. 441–452. Stanford University, Stanford, CA, USA (2009)

Westfälische Hochschule (ed). Bachelor Sustainable Engineering & Management (2024). https://www.w-hs.de/sustainable-engineering-boh/. Accessed 31 Jan 2024

Zhao, S.: Integriertes Management von ökonomischen und ökologischen Produkteigenschaften: Eine EcoDesign-Methodik für kleine und mittlere Unternehmen (Zugl.: Darmstadt, Techn. Univ., Diss., 2012), Düsseldorf, VDI-Verl (2013)

Author Index

K. Gericke et al. (Eds.): DSC 2022/2023, PTT, p. 249, 2026.
https://doi.org/10.1007/978-3-032-16278-6

MIX
Papier aus verantwortungsvollen Quellen
Paper from responsible sources
FSC® C105338

If you have any concerns about our products,
you can contact us on
ProductSafety@springernature.com

In case Publisher is established outside the EU,
the EU authorized representative is:
Springer Nature Customer Service Center GmbH
Europaplatz 3, 69115 Heidelberg, Germany

Printed by Libri Plureos GmbH
in Hamburg, Germany